Express Your Creativity
with Adobe Express

Create stunning graphics, captivating videos, and impressive
web pages without any coding skills

Rosie Sue

BIRMINGHAM—MUMBAI

Express Your Creativity with Adobe Express

Group Product Manager: Rohit Rajkumar
Publishing Product Manager: Nitin Nainani
Senior Content Development Editor: Feza Shaikh
Technical Editor: Simran Udasi
Copy Editor: Safis Editing
Project Coordinator: Manthan Patel
Proofreader: Safis Editing
Indexer: Tejal Daruwale Soni
Production Designer: Ponraj Dhandapani
Marketing Coordinators: Anamika Singh, Namita Velgekar, and Nivedita Pandey

First published: July 2023

Production reference: 1140623

Published by Packt Publishing Ltd.
Livery Place
35 Livery Street
Birmingham
B3 2PB, UK.

ISBN 978-1-80323-774-9

www.packtpub.com

I extend my heartfelt appreciation to the memory of my father, Leonard Sue, and my mother, June Sue, for the immense sacrifices they made that have enabled me to lead a blessed life. Their unwavering support has been the foundation of my success, and I am forever grateful for their selflessness and devotion.

I also wish to express my gratitude to my partner, Leonardo Alonso, for his unwavering patience during my long hours of writing, often locked away while exclaiming, "Just one more chapter, and I'll be free!" Your creative flair and unwavering dedication to your craft inspire me every day, and I am grateful to have you as my biggest champion and cheerleader.

To everyone who has been a part of my professional and personal journey, I offer my sincerest appreciation. The guidance, mentorship, and support you have provided have been instrumental in shaping the trajectory of my life and career, and I am blessed to have crossed paths with each of you.

– Rosie Sue

Contributors

About the author

Rosie Sue is a talented, highly skilled, and accomplished content creator, instructor, podcaster, and aspiring artist based in Auckland, New Zealand. She has built a successful YouTube channel featuring technical educational tutorials, which have garnered over 450,000 views to date. Rosie's expertise and dedication to teaching have also earned her recognition as a top 3% instructor on Udemy, where she has published two online courses with over 1,500 student enrolments.

In addition to her bachelor's degree in communications studies from AUT University, Rosie earned a postgraduate certificate in design from Media Design School, where she was honored as the GPA Award winner for achieving the highest GPA in her graduating class. Rosie's academic achievements, combined with her natural creativity and passion for design, have made her a valuable asset to her current employer, Adobe, where she works as a solutions consultant. In this role, Rosie has used her expertise to help some of the country's largest organizations foster creativity and improve their design workflows for the past six years.

In 2022, Rosie was a member of the selection panel for the New Zealand Youth Film Festival (NZYFF), and she was also invited to appear as a guest on New Zealand's national breakfast television show, *Breakfast*, on TVNZ, alongside the founder. These experiences have allowed Rosie to further expand her professional network and share her insights and expertise with a wider audience.

Rosie's list of accomplishments also includes the title of published author, with her first book on Adobe Express being the first of many (we hope). Rosie is constantly seeking new opportunities to grow and develop her skills, and her hard work and dedication have paid off in the form of a diverse and impressive portfolio of work.

To learn more about Rosie and stay up to date on her latest projects, be sure to visit her website at `https://rosiesue.co.nz/`.

About the reviewer

Julius Atienza Aala is a graphic designer, Adobe Express ambassador, and creative educator based in the Philippines. His firm passion for art fuels his relentless pursuit of excellence. He continually seeks to enhance his knowledge and skill set, immersing himself in diverse courses and job opportunities. His unwavering dedication has shaped him into a true master of his craft.

Driven by a deep desire to share his expertise, he has emerged as a beacon of guidance in the field of education. As a graphic designer/educator, he has taken the stage to deliver captivating talks across varied sectors. Through these engagements, he imparts invaluable wisdom on graphic design software, unravels the fundamentals of this captivating art form, and highlights the importance of digital literacy in today's era.

Table of Contents

Part 2 – Create Graphics with Adobe Express

3

4

5

6

Part 4 – Create a Video with Adobe Express

11

12

13

Preface

Welcome to this ultimate and comprehensive guide that will teach you how to create visually stunning content using Adobe Express. Whether you're a social media manager, small business owner, or someone interested in upskilling in graphic design, this book is your go-to resource for creating beautiful and captivating content without requiring any design skills or experience.

Adobe Express is an easy-to-use platform that offers thousands of pre-designed templates to choose from, which you can personalize to your brand. With Adobe Express available on both web and mobile devices, creating engaging content has never been easier.

This book is designed to take you through every function available on Adobe Express, providing you with practical exercises to ensure that you can put what you learn into practice. Throughout this book, you will gain comprehensive knowledge on a range of topics, from formatting typography to creating animations and captivating videos, as well as designing beautiful landing pages and social media graphics that are sure to generate engagement.

Whether you are an absolute beginner or an experienced graphic designer, this book will equip you with the knowledge and skills needed to create stunning content. By the end of this book, you will have mastered Adobe Express and be able to create stunning social media graphics, beautiful landing pages, and captivating videos with ease.

Who this book is for

Irrespective of your profession, be it a marketer, content creator, aspiring designer, or entrepreneur, the need to generate content for social media, brochures, web pages, and marketing materials is a constant requirement.

This book will enable you to design without any prior experience with the software or graphic design. This book will guide you through the process and enable you to create stunning graphics, web pages, and videos.

What this book covers

Chapter 1, *A Brief Introduction to Adobe Express*, teaches you how to create stunning graphics, animations, landing pages, and videos for social media and other marketing purposes using Adobe Express.

Chapter 2, Express Your Brand – How to Create Your Brand, shows you how to elevate your visual communication with Adobe Express by easily applying your brand's font, logo, and color scheme with just a single tap, ensuring consistency across all your content.

Chapter 3, Creating Expressive Content Starting with a Template from Adobe Express, explores how to easily navigate and customize thousands of professionally designed templates in Adobe Express, including social media posts, logos, and brochures.

Chapter 4, Level Up Your Social Media Posts with Adobe Express, discusses how to customize your social media posts, including resizing and changing fonts, and adding images and icons.

Chapter 5, Animating Text and Images with Adobe Express, delves into how to easily create and export professional-looking text and image animations to level up your social media game and capture your audience's attention.

Chapter 6, Editing Images Using Quick Actions, examines how to remove image backgrounds, resize and crop images, and convert between PNG and JPG formats.

Chapter 7, Polishing PDFs Using Quick Actions, explores tasks such as converting a PDF into a JPG and vice versa, and organizing and combining PDF files, which are just a few of the available PDF editing tools you can utilize in Adobe Express. The powerful features of Adobe's world-standard PDF software, Acrobat, can now be accessed inside Adobe Express, under **Quick Actions**.

Chapter 8, Put Your Skills to Practice with Adobe Express, includes three practical exercises to help you learn Express, design techniques, and ways to use its various functions. You will create an Instagram story, a marketing campaign, and an event poster, using layers, design assets, and images to create dynamic designs that will impress professional graphic designers.

Chapter 9, Building a Web Page with Adobe Express, teaches you how to create a professional one-page website without coding, featuring images, videos, and formatted text, and making it shareable for free.

Chapter 10, Mini Projects – Creating Your Own Web Page(s) with Adobe Express, guides you through creating web pages without coding skills, with practical exercises and examples provided. Discover the versatility of web pages and learn how to create engaging and effective pages for various purposes, all hosted on Adobe's platform and easily shareable with others.

Chapter 11, Creating and Editing Videos, discusses how to create and edit videos using Adobe Express. You will learn how to add videos, images, icons, text, and music to create a polished video, adjust audio, record voice-overs, and share your final video.

Chapter 12, Polishing Videos Using Quick Actions, explores video Quick Actions in Express, powered by Adobe Premiere Pro, allowing you to effortlessly enhance your videos with features such as resizing, converting to GIFs, cropping, adjusting speed, and converting to MP4, in just a few minutes.

Chapter 13, Scheduling Content in Adobe Express, explores how to automate the process of scheduling and posting your content on multiple social media platforms using the Content Scheduler in Express, allowing you to plan and batch-process your content ahead of time and build a social media strategy.

To get the most out of this book

Before you begin using Express on your browser or the app, there are a few things you should keep in mind. First, ensure that you have access to a modern browser that can support the application. Some older browsers may not be able to handle the features and functionality of Express, so it's essential to ensure that your browser is up to date.

Adobe Express on the web:

Software/hardware covered in the book	Operating system requirements
Operating systems	Windows: Version 8.1 or later macOS: Version 10.13 or later Chromebook
Web browsers	Chrome, Firefox, Safari, and Edge Note – JavaScript must be enabled
Memory requirements	Minimum 4 GB memory

Adobe Express on iOS and Android:

Software/hardware covered in the book	Operating system requirements
Operating systems	Minimum requirement
iOS	iOS 14 or later
Android	Android 9.0 Pie or later

A note to users – an internet connection, Adobe ID, and acceptance of the Adobe Terms of Use are required (www.adobe.com/go/terms) to activate and use this product. This product may integrate with or allow access to certain Adobe or third-party hosted online services. Users must be 13 or older to register for an individual Adobe ID. Adobe products and services may not be available in all countries or languages and may be subject to change or discontinuation without notice.

Download the color images

We also provide a PDF file that has color images of the screenshots and diagrams used in this book. You can download it here: https://packt.link/mHj53.

Conventions used

There are a number of text conventions used throughout this book.

`Code in text`: Indicates code words in text, database table names, folder names, filenames, file extensions, pathnames, dummy URLs, user input, and Twitter handles. Here is an example: "Mount the downloaded `WebStorm-10*.dmg` disk image file as another disk in your system."

Bold: Indicates a new term, an important word, or words that you see on screen. For instance, words in menus or dialog boxes appear in **bold**. Here is an example: "Click on the **Add your image** option and add an image from Adobe Stock."

Tips or important notes
Appear like this.

Get in touch

Feedback from our readers is always welcome.

General feedback: If you have questions about any aspect of this book, email us at `customercare@packtpub.com` and mention the book title in the subject of your message.

Errata: Although we have taken every care to ensure the accuracy of our content, mistakes do happen. If you have found a mistake in this book, we would be grateful if you would report this to us. Please visit `www.packtpub.com/support/errata` and fill in the form.

Piracy: If you come across any illegal copies of our works in any form on the internet, we would be grateful if you would provide us with the location address or website name. Please contact us at `copyright@packt.com` with a link to the material.

If you are interested in becoming an author: If there is a topic that you have expertise in and you are interested in either writing or contributing to a book, please visit `authors.packtpub.com`.

Share Your Thoughts

Once you've read, we'd love to hear your thoughts! Scan the QR code below to go straight to the Amazon review page for this book and share your feedback.

https://packt.link/r/1803237740

Your review is important to us and the tech community and will help us make sure we're delivering excellent quality content.

Download a free PDF copy of this book

Thanks for purchasing this book!

Do you like to read on the go but are unable to carry your print books everywhere?

Is your eBook purchase not compatible with the device of your choice?

Don't worry, now with every Packt book you get a DRM-free PDF version of that book at no cost.

Read anywhere, any place, on any device. Search, copy, and paste code from your favorite technical books directly into your application.

The perks don't stop there, you can get exclusive access to discounts, newsletters, and great free content in your inbox daily

Follow these simple steps to get the benefits:

1. Scan the QR code or visit the link below

https://packt.link/free-ebook/9781803237749

2. Submit your proof of purchase
3. That's it! We'll send your free PDF and other benefits to your email directly

Part 1 – A Brief Introduction to Adobe Express

In this part, you will discover how to unleash your creativity with Adobe Express, an intuitive platform that allows you to create stunning graphics, animations, landing pages, and videos. Whether you're a marketer, content creator, designer, or entrepreneur, the ability to create captivating content is crucial for your success. Adobe Express makes it easy to create content for various platforms, including social media, brochures, web pages, and marketing collateral, without any coding skills required. Additionally, you will also learn how to manage your brand style by uploading your assets and ensuring consistency in all visual communications. Get ready to unleash your creativity with Adobe Express.

This part has the following chapters:

- *Chapter 1, A Brief Introduction to Adobe Express*
- *Chapter 2, Express Your Brand – How to Create Your Brand*

1

A Brief Introduction to Adobe Express

In this book, you will learn how to create stunning social media graphics, thumb-stopping animations, beautiful landing pages, and captivating videos. Additionally, you will be introduced to the power of Quick Actions in Express, which will enable you to swiftly edit videos and effortlessly create PDFs. You will discover ways to integrate your brand with Express and leverage the social media content scheduler. These are just some of the many techniques that you will master in this book. Enjoy the flexibility of creating beautiful content either in the browser or on your mobile device using the Adobe Express apps.

Regardless of your professional background, whether you're a marketer, a content creator, an educator, a budding designer, or an entrepreneur, it is crucial to consistently generate content. Express is an entry-level tool that revolutionizes the design landscape, ensuring accessibility to everyone. By democratizing design, Adobe Express empowers individuals with any design expertise to effortlessly create visually beautiful content.

Learning and putting what you've learned into practice has never been easier, and you can accomplish something no matter which chapter you read.

Whether you want to start from scratch or select from a curated selection of templates, you can create beautiful content within minutes. You can create social media graphics, social media posts, Instagram stories, YouTube thumbnails, event posters, business cards, flyers, logos, and much more with Adobe Express.

With Adobe Express, you can create web pages without any coding skills. Web pages can be created within minutes, and they are hosted on Adobe. You can add hyperlinks, images, and videos to these web pages. Expand your creative output for your brand by creating web pages for your product or business; create a marketing splash page to advertise an upcoming event; or launch an announcement with a call to action with an Adobe Express web page.

Go a step further and create videos and animations with Adobe Express. Editing a video using Express doesn't come with a steep learning curve. You can effortlessly incorporate images and videos, and utilize the app to trim, crop, and resize your videos. Enhance the impact of your videos by incorporating a soundtrack from the extensive library of free music available in Express. Additionally, recording voiceovers in Express is a simple and straightforward process.

We'll cover the following topics in this chapter:

- What is Adobe Express?
- How can I access Adobe Express?
- Creating an account on Adobe Express
- Adobe Express Premium versus free versions

By the end of the chapter, you will be able to create an account in Adobe Express, and you will know the difference between the free and Premium versions. You can decide to pay for the premium version at any point.

Technical requirements

You will need a modern browser, such as Google Chrome. Additionally, you can download the Express app, which is slightly functionally different from the browser version.

What is Adobe Express?

Adobe Express is a web and mobile app that allows you to easily create stunning graphics and content without having any graphic design skills. You can choose from thousands of templates and personalize and remix the content to create your own eye-catching content.

Examples of the type of content you can create

Here are some examples of content that you can create in Adobe Express:

- **Social media content**

  You can create social media content quickly and easily by using one of the hundreds of templates provided by Adobe Express.

Figure 1.1 – Example of a social media post

- **A logo**

  With Adobe Express, you can create logos by using the logo builder. Adobe Express makes it easy for you to create a customized logo from scratch with their easy step-by-step guide.

Figure 1.2 – An example of a logo

- **Graphics for Instagram/Facebook stories and covers for reels and TikTok**

  Create captivating, thumb-stopping content for your Instagram and Facebook stories, and covers for Instagram/Facebook Reels and TikTok. Choose from one of the hundreds of modern templates available in Adobe Express.

Figure 1.3 – An example of an event graphic created from a template, which can be animated and used for your Instagram stories

- **Create a business card**

  With Adobe Express, you can create a professional business card for your business.

Figure 1.4 – An example of a business card

- **Presentation graphics**

  Forget boring PowerPoint presentations! With Adobe Express, you can easily create professional presentation graphics by choosing from hundreds of templates.

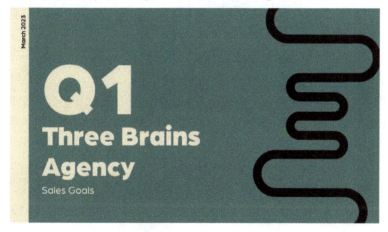

Figure 1.5 – An example of a business presentation slide

- **A menu**

You can create unique and modern menus quickly and easily using a template with Adobe Express.

Figure 1.6 – An example of a menu

- **An invitation**

You can create birthday invitations, wedding invitations, baby shower invitations, and much more. With Adobe Express, choose from hundreds of stunning templates for your next party invitation.

Figure 1.7 – An example of a wedding invitation

- **A web page**

    Need a website or marketing splash page for your business but don't know how to code? Look no further! With Adobe Express, you can quickly and easily create professional web pages.

Figure 1.8 – An example of a web page created in Adobe Express

These are just a handful of examples of the type of content you can create in minutes with Adobe Express. There is no limit to what you can create with Adobe Express.

How can I access Adobe Express?

Adobe Express can be accessed on the browser at the following link:

```
https://express.adobe.com/
```

There is no need to download any software. Adobe Express can be accessed through a browser, so no matter where you are, you can simply navigate to the URL and start creating.

Adobe Express is also available as an app on iOS and Android. You can start creating on your iPhone, iPad, or Android device.

Adobe Express for iOS

To start Adobe Express on iOS, follow these steps:

1. Visit the Apple App Store and download the Adobe Express app.
2. On your iPad or iPhone, tap on the Adobe Express app.
3. Sign in with your Adobe ID, Apple ID, Facebook account, or Google account (refer to the *Creating an account in Adobe Express* section to learn more).
4. Once you have signed in, you can start creating!

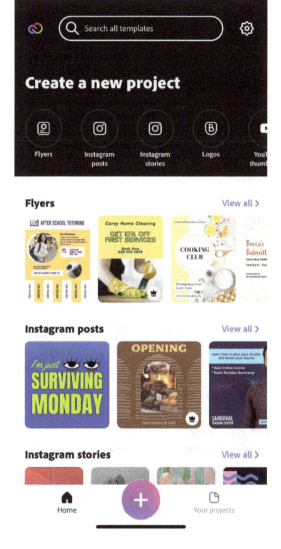

Figure 1.9 – Adobe Express app home page

Adobe Express for Android

To start Adobe Express on Android, follow these steps:

1. Visit the Google Play store and download the Adobe Express app.
2. On your Android device, tap the Adobe Express app.

3.   Sign in with your Adobe ID, Facebook account, or Google account (refer to the next section to learn more).

4.   Once you have signed in, you can start creating!

Now that we have learned how to access Adobe Express on the browser and through the mobile apps, in the next section, we will learn how to create an account.

Creating an account in Adobe Express

Whether you are using the free or Premium version of Adobe Express, you can create an Adobe ID.

If you are using the free version of Adobe Express, you do not need to create an Adobe ID. You can sign up to use Adobe Express with your Facebook account, Google account, or Apple ID.

To create an account in Adobe Express, follow these steps:

1.   Navigate to `https://express.adobe.com/`.

2.   Click on **Sign Up**:

Figure 1.10 – Sign Up button

3.   Next, select **Log in with Adobe ID**:

Figure 1.11 – Login page for Adobe Express on the web

4. Next, on the **Sign in** page, click on **Create an account**:

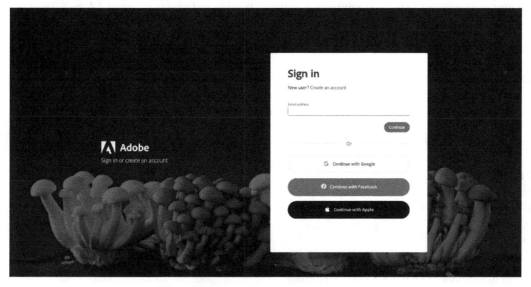

Figure 1.12 – Sign-in page for Adobe Express on the web

5. Fill out your details and select **Create Account**:

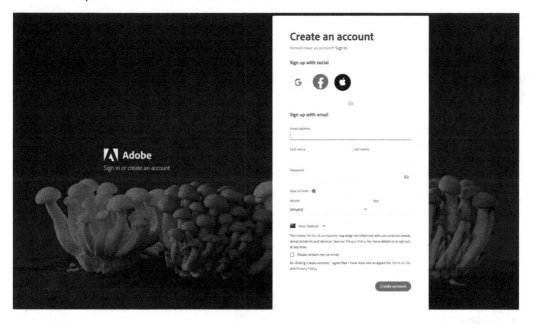

Figure 1.13 – Create an account page for Adobe Express on the web

Now that you have created an account, you can use your Adobe ID to log in to Adobe Express. In the next section, you will learn the difference between the free and paid versions of Adobe Express.

Adobe Express Premium versus free versions

You can choose to use the free plan or pay for an Adobe Express membership. With the free plan, you can access limited content and features within the Adobe Express app. However, with the Premium version, you can access thousands of premium templates and assets.

Let's look at each of the versions in detail.

Free plan

The free plan includes all the core features and has some limitations:

- 2 GB cloud storage.
- Access thousands of templates, Adobe Fonts, and design assets. A limited collection of royalty-free Adobe Stock free collection photos.
- Basic editing and photo effects including remove background and animate.
- Create on both web and mobile.

Premium plan

You can pay for a subscription to unlock the premium features of Adobe Express.

Features include the following:

- 100 GB of cloud storage
- Access to all the premium templates, design assets, and more than 160 million roaylty-free Adobe Stock collection photos.
- Collaborate on and share projects with your team
- Manage templates and assets with Creative Cloud Libraries
- Access to all the premium editing features, including resizing images and refining cutouts
- Access to more than 20,000 Adobe Fonts and character types (curved type, font pairs, and grids)
- Create your own brand (branding, logo, colors, and fonts)
- Create PDFs and export them other file types
- Create on web and mobile

You will also get access to the following apps when you subscribe to an Adobe Express membership:

- The Adobe Express app (web and mobile)
- Adobe Spark Page (mobile)
- Adobe Spark Video (mobile)
- Adobe Photoshop Express (mobile)
- Adobe Premiere Rush (desktop and mobile)

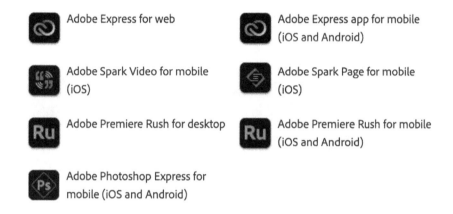

Figure 1.14 – A premium subscription includes all these additional Adobe apps

Summary

Throughout this chapter, you gained insights into the wide array of content you can design with Adobe Express. You were introduced to different ways of accessing Express: via the browser or the mobile apps. As part of this process, you have created an Adobe ID, which will serve as your login credentials. As explained in this chapter, you have the option to upgrade from the free version to the Premium version at any time.

In the upcoming chapter, you will delve into the creation and integration of your brand within Adobe Express. You will learn how to create a style guide in Adobe Express, encompassing elements such as your logo, font selection, and color scheme. These elements will ensure you have a consistent and cohesive visual identity across all the content you create using Express.

2

Express Your Brand – How to Create Your Brand

The chances are you already have a brand/corporate guideline encompassing a specific font, a logo, and perhaps even a color scheme. With Adobe Express, you can upload these assets and set up your brand within the platform. With the magic of Adobe, you can effortlessly apply your logo, font, and color scheme to all your projects with just a single tap. This ensures all your content maintains consistency and elevates your visual communication to your audience, whoever they may be.

With your style guide uploaded as a brand on Adobe Express, you can do the following:

- Easily add your logo to your content with just one click
- Apply your brand colors to your content with just one click
- Maintain brand consistency by applying your company font to your content

Managing your brand style in Adobe Express is simple; you only need to upload your brand assets once. Additionally, you can also create multiple brands in Adobe Express (*Figure 2.1*). So, if you have an agency, for example, and you have multiple clients, you can create multiple brands in Adobe Express.

Here are some of the fictitious brands I have created in Adobe Express. As you can see, I have created three different brands: Isabella's Cantina, Brand x, and Delicious Treats.

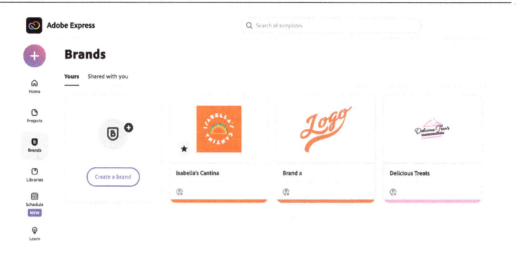

Figure 2.1 – Three different brands in Adobe Express

We will cover the following topics in this chapter:

- Getting started with creating a brand

- How to upload a logo and select brand colors

- How to select or upload a font and decide on a brand name

By the end of this chapter, you will possess the skills to establish your brand within Express. You will be able to create a brand from scratch and learn the process of uploading or selecting a font, uploading your logo, and naming your brand within the platform. With your newfound knowledge, you can successfully create a unique brand identity and maintain consistent branding across all your content creation.

Get started with creating a brand

To get started, follow these steps on the browser:

1.   Navigate to the Adobe Express home page, https://express.adobe.com/, in your browser:

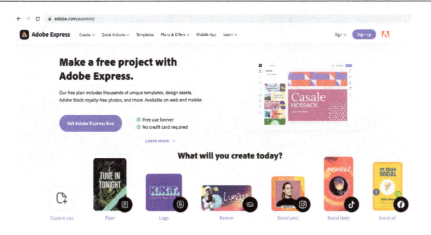

Figure 2.2 – Adobe Express in the browser

2.   Select **Brands**, located to the far left of your screen.

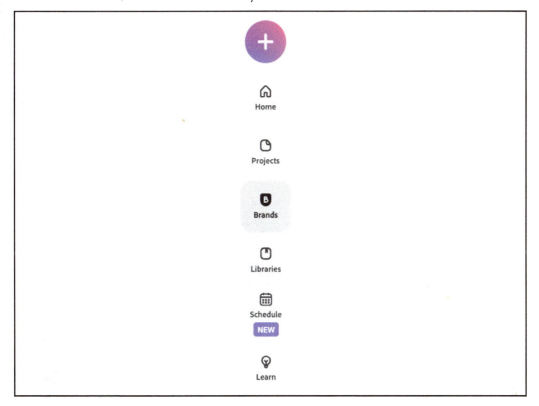

Figure 2.3 – Menu on the left

3.   When you navigate to the **Brands** tab, you will be taken to the **Brands** home page. By default, you will be automatically directed to the **Yours** tab. This tab is where Adobe Express will host all the brands you create.

4.   You can also navigate to the **Shared with you** tab. Here, you will be able to see brands others have given you access to. When someone grants you access to a brand, you will have the ability to either view or edit the brand, depending on the permission level they have granted you access to. For example, if you want to collaborate with a team, only one person will need to create a brand. They can then share the brand with the rest of the team to use.

5.   To get started and create your first brand, click on **Create a brand**.

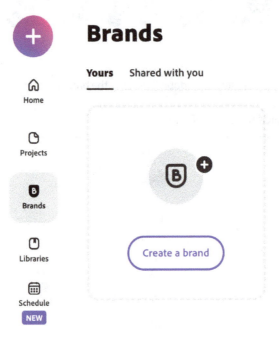

Figure 2.4 – Brands menu option

6.   Once you click on **Create a brand**, Adobe Express will automatically load the three elements you will need to upload: your logo, color, and font.

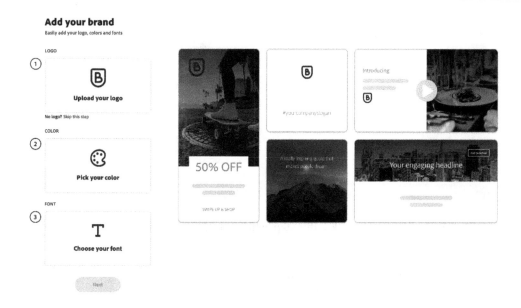

Figure 2.5 – Interface to upload your brand elements

Having acquired the basics of creating a brand, we are now ready to delve into the next section. In the upcoming section, you will get an insight into the process of uploading your logo and selecting the brand colors.

Uploading a logo and selecting brand colors

In this section, you will learn how to start uploading important brand assets that make your brand unique. You will learn how to upload your logo and how to either pick a color from your logo or input your hex color code.

Continuing from the previous section, you can follow these steps to upload your logo and choose a color for your brand:

1. Navigate to **LOGO,** hover your cursor over the rectangle, and click on **Upload your logo**. When you hover your cursor over the rectangle, Adobe Express will tell you the file formats they support are JPG and PNG. You can upload a JPG or PNG copy of your logo.

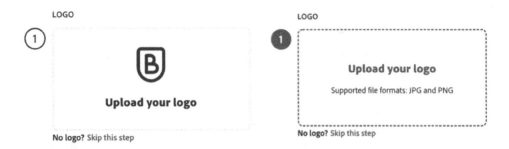

Figure 2.6 – Upload your logo

2.   If you don't have a logo, click on **Skip this step**, which is located at the bottom of the rectangle.

Figure 2.7 – Option to click on Skip this step when prompted to upload your logo

3.   If you click on **Skip this step**, Adobe Express will display a placeholder logo with the letter **B**, as shown in the following screenshot.

4.   Adobe Express will also give you the option to **Try our logo maker**, which is a hyperlink located at the bottom of the rectangle.

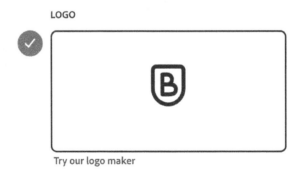

Figure 2.8 – Logo creation tool in Adobe Express

5. To create a logo using Adobe Express, follow these steps:

I. Once you click on **Try our logo maker**, you will be directed to the **Logo Maker** page, which opens in a new browser. Here, Adobe Express will ask you for the information it requires from you. In the following screenshot, you can see the information required is the type of business, the name of your business, and your slogan:

Figure 2.9 – Question prompts from Adobe Express

II. For example, in the following screenshot, I have created an indoor plant company:

Figure 2.10 – Answer the questions for your business information

III.   Once you click **Next**, Adobe Express will ask you to choose a style. Here, they provide you with four options to describe your aesthetic: **Bold**, **Elegant**, **Modern**, or **Decorative**.

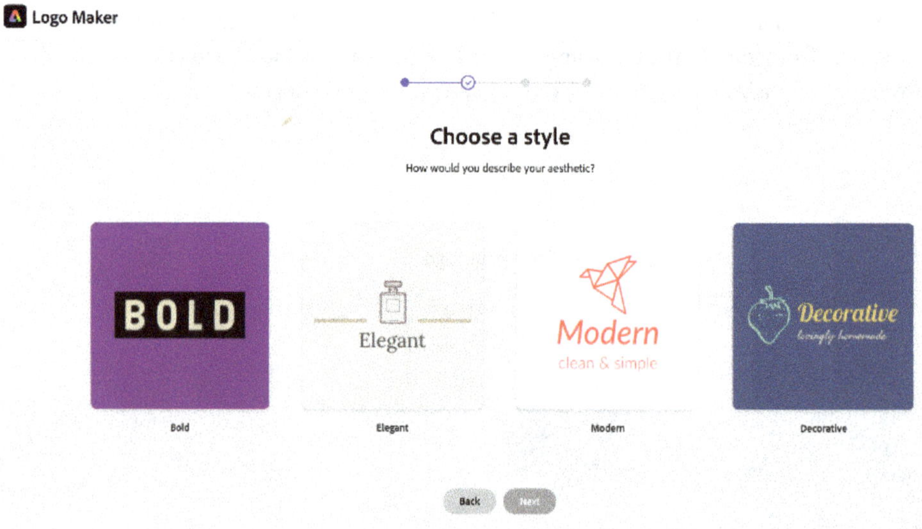

Figure 2.11 – Select from Bold, Elegant, Modern, and Decorative aesthetics

IV.   Select **Modern**, then click **Next**.

V.   Next, Adobe Express will ask you to choose an icon. Type in a keyword to find a suitable icon.

Figure 2.12 – Input a keyword and search for an icon

VI.   Select whichever icon you want and click **Next**.

VII. Adobe Express will then display the logos it has created. These are only some of the results Adobe Express has come up with. Amazing!

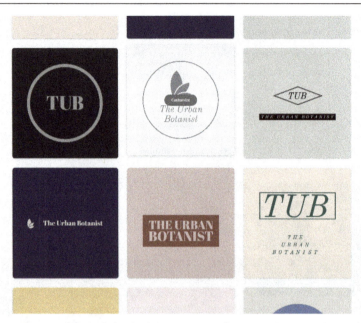

Figure 2.13 – With the magic of Adobe, Express will give you logos to choose from

I.   When you reach the bottom of the page, Express will present you options to view more logo variations, along with the option to choose a new symbol. Click on any of the styles to view more logo options.

Figure 2.14 – Choose a different style or change your symbol

II.   Click on the logo of your choice. Here is the logo I selected for the fictitious company I created:

Figure 2.15 – The logo I picked in this example

III.   To the right of the logo, you can click on **Color**. Adobe Express will shuffle through a range of colour combinations for you to choose from. You can also click on **Font** to shuffle through some font options.

IV.   Next, click on **Download**. Adobe Express will download a ZIP file, which will contain a copy of your logo in the colour you picked, a copy of your logo in black and white, and also a transparent copy.

1.   After creating your logo, you can navigate back to the browser to continue uploading your brand. Refer to *step 1* to upload your newly created logo; otherwise, you can jump straight to *step 7* if you have already uploaded your own logo.

2.   Next, we will add your color to your brand. Click on **Pick your color**.

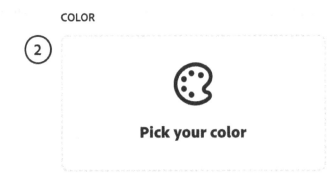

Figure 2.16 – Pick a color

3.   When you hover your cursor over the color dialog box, Adobe Express displays a tip that states the following: **Select the color from your logo or choose your own color**.

Figure 2.17 – Hover state will display a guide

4.   Once you click **Pick your color**, Adobe Express will automatically display a color that it has extracted from your logo. It will also display the hex code of the color from your logo.

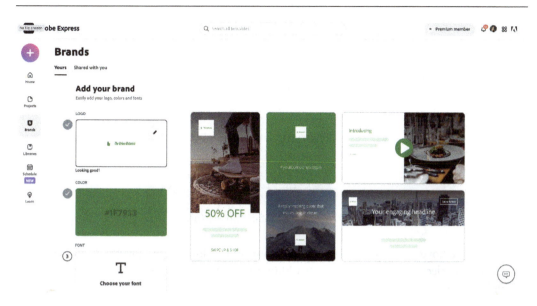

Figure 2.18 – Color wheel

In this section, we uploaded and/or created a logo. You also learned how to add a color to your brand. In the next section, we will finish creating our brand by choosing a font and, finally, adding the name of your brand in Adobe Express.

How to select or upload a font and decide on a brand name

In this section, you will learn how to choose a default font from Adobe Express or upload your own font. You will also learn how to name your brand and save your brand in Adobe Express.

To get started, follow these steps:

1. To add a font to your brand, navigate to **Choose your font**.

Figure 2.19 – Choose a font in Adobe Express

2. When you hover over this option, **Choose your font**, Adobe Express will tell you that you can either select from a library of fonts within Adobe Express, or you have the option to also upload your own.

Figure 2.20 – Hovering over the FONT box will display a suggestion guide

3. Once you click on **Choose your font**, you have the option to click on **Add your fonts** to upload your own font. Alternatively, you can scroll through the library of fonts provided by Adobe Express.

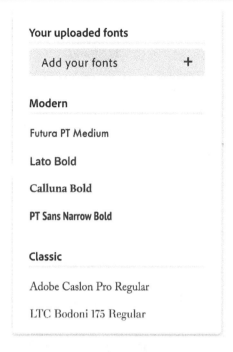

Figure 2.21 – Selecting an Adobe Font or uploading your own font

4.   I have selected the font called **Museo Sans Rounded 500**.

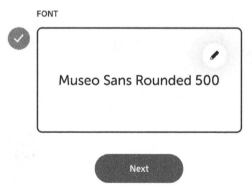

Figure 2.22 – I selected the Museo Sans Rounded 500 font for this example

To create your brand, the final step is to select a brand name.

Follow these steps to give your brand a name:

1. Click on the **Next** button, which is below the **FONT** option. Adobe Express will display a popup, asking you to input the brand name.

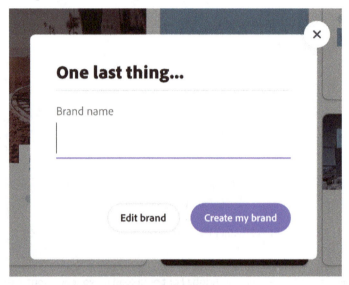

Figure 2.23 – Input the name of your brand

2. I have typed in `The Urban Botanist`. You can add your own brand name here.

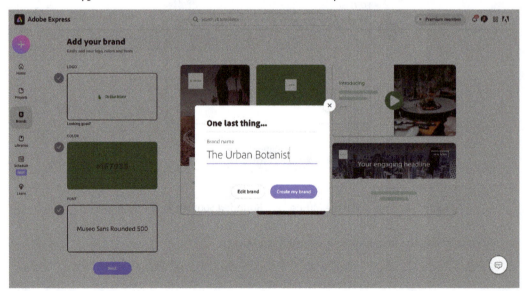

Figure 2.24 – I've called my brand The Urban Botanist for this example

3.   Next, click on the **Create my brand** button. Adobe Express will take less than a minute to create your brand. Here is a screenshot of the placeholder image.

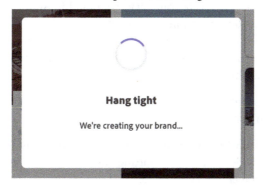

Figure 2.25 – Adobe Express will display the progress of creating your brand

4.   That's it! Adobe Express will now direct you to your newly created brand! You have the option to add other variations of your logo, more colors, and more fonts if you wish.

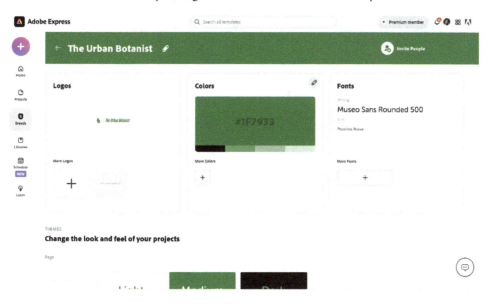

Figure 2.26 – The brand I created in this chapter

Summary

In this chapter, you learned how to create your very own brand in Adobe Express. By establishing your brand in Adobe Express, you can guarantee a consistent and cohesive visual representation in all the content you create within the platform. Uploading your brand in Express allows you to align your corporate identity with every piece of content you create. Customized branding on all your content will help your audience remember your brand.

Furthermore, this chapter also introduced you to the logo maker within Adobe Express. With Adobe's logo builder, you can design a stunning logo within minutes.

In the upcoming chapter, you will start creating content using pre-built templates within Adobe Express. These templates provide a convenient starting point for your content, allowing you to craft visually appealing designs with efficiency and ease.

Part 2 – Create Graphics with Adobe Express

Adobe Express offers a variety of tools and features to help you create expressive content. You can easily navigate and customize thousands of professionally designed templates, including social media posts, logos, and brochures. Additionally, you can level up your social media posts by customizing them, including resizing, changing fonts, and adding images and icons. Animating text and images is also easy with Adobe Express, allowing you to create and export professional-looking text and image animations to capture your audience's attention. Furthermore, you can edit images using Quick Actions, such as removing image backgrounds, resizing, cropping, and converting between PNG and JPG formats. Polishing PDFs is also made easy with Adobe's world-standard PDF software, Acrobat, which can be accessed right inside Adobe Express. Finally, mini projects are available to help you practice your skills, including creating an Instagram story, marketing campaign, and event poster, using layers, design assets, and images to impress professional graphic designers.

This part has the following chapters:

- *Chapter 3, Creating Expressive Content Starting with a Template from Adobe Express*
- *Chapter 4, Level Up Your Social Media Posts with Adobe Express*
- *Chapter 5, Animating Text and Images with Adobe Express*
- *Chapter 6, Editing Images Using Quick Actions*
- *Chapter 7, Polishing PDFs Using Quick Actions*
- *Chapter 8, Put Your Skills to Practice with Adobe Express*

3

Creating Expressive Content Starting with a Template from Adobe Express

In this chapter, you will be introduced to the process of remixing and repurposing templates in Express. You will learn how to seamlessly navigate through thousands of templates, explore various categories, and effectively search for specific templates based on your needs. You will see how to browse through diverse categories such as brochures, business cards, presentation graphics, social media posts, logos, and much more.

You will also learn how to filter templates by topics such as social media lifestyle business and much more, along with filtering for free, premium, or animated content.

With Express, you have access to thousands of professionally designed templates, so regardless of your skill level, you can customize these templates and produce professional and polished content.

We will cover the following topics in this chapter:

- How to create a social media post from a template
- How to create a logo from a template
- How to create a brochure from a template

By the end of this chapter, you will have honed your skills to create any content by simply starting with a template provided by Adobe Express. You will be able to create stunning social media posts and eye-catching content, which you can use on your Instagram Stories, Instagram Reels, LinkedIn posts, and covers for your TikTok. Additionally, you will acquire the knowledge to create a logo from a premade template, customizing it to make it your own. And finally, you will be able to create a professional brochure to advertise your business using a template from Adobe Express.

How to create a social media post from a template

To get started, follow these steps on the browser:

1. Navigate to `https://express.adobe.com/`.

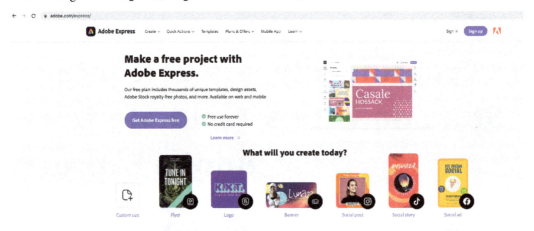

Figure 3.1 – Home page of Adobe Express on the browser

2. Navigate to the + icon, located at the top-left corner of the page, and click it once.

Figure 3.2 – Click the + icon

3. When you click on the + icon, you will be presented with options to either create new content or get started with quick actions.

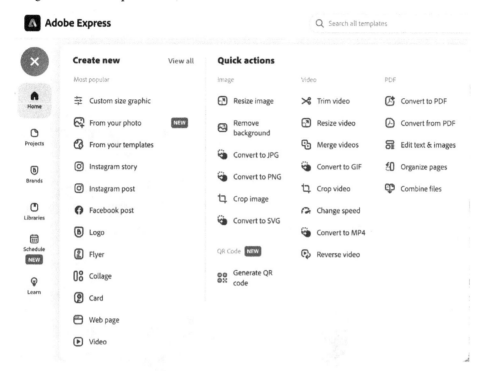

Figure 3.3 – TheCreate new and Quick actions options in Adobe Express

4. Click on **Instagram story**.

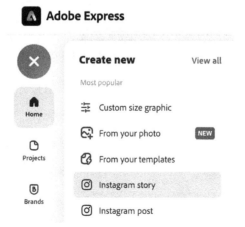

Figure 3.4 – Template options, including Instagram story and Instagram post

5. When you click on **Instagram story**, you are redirected to a page displaying all the templates available for Instagram stories. There are seven categories: **Business**, **Food and drink**, **Inspirational**, **Interactive**, **Sales and promotions**, **Travel**, and **Minimal**. Adobe Express allows you to select a category and will filter the templates accordingly.

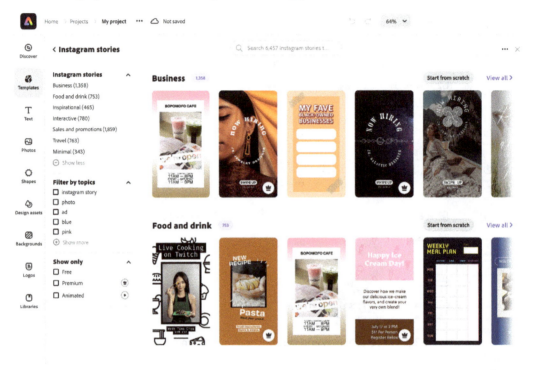

Figure 3.5 – Instagram stories templates

6. Select the **Business** category to view all the templates available in this category. Select a template you like by simply clicking on the selected template thumbnail.

7. Once you have selected a template, you can now edit and customize it. On the left-hand pane, the following options are available: **Discover**, **Templates**, **Text**, **Photos**, **Shapes**, **Design assets**, **Backgrounds**, **Logos**, and **Libraries**. On the right-hand pane, the following options are available: **Colors**, **Animation**, **Layout**, **Resize**, and **Design**.

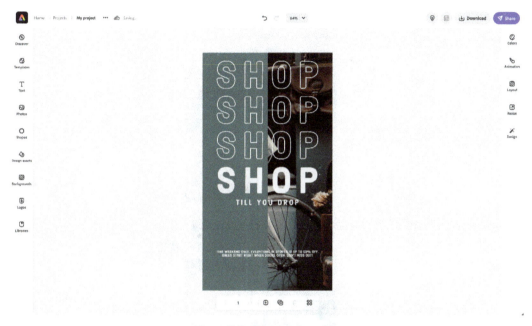

Figure 3.6 – Business template

8. To edit the template, simply click on any of the elements, such as the text or image.

9. When you double-click on the image, the **Photos** option window will pop out from the sidebar and present you with photos from Adobe Stock.

Figure 3.7 – You can find high-quality images from Adobe Stock right inside Adobe Express

10. Alternatively, you can click on the **Upload photo** button to upload your own images.

11. Additionally, if you click on the ellipsis (three dotted lines), you will be presented with more options to bring in an image from other third-party photo depositories, including Dropbox, Google Drive, and Google Photos.

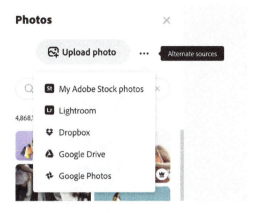

Figure 3.8 – When you click on the ellipsis, Adobe Express allows you
to bring in photos from third-party photo depositories

12. To search for images on Adobe Stock, simply search for a keyword, such as macaroons.

Figure 3.9 – Type in a keyword to search for images in Adobe Stock

13. Simply click on the image of your choice. Adobe Express will place the image on your template. You can drag the image with your cursor to reposition it.

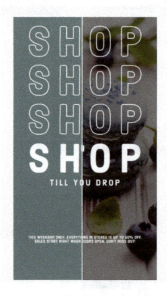

Figure 3.10 – Added an image from Adobe Stock to replace the existing image on the template

14. Next, to edit the text, simply click on the existing text on the template. On the right-hand pane, the **Edit text** flyout menu will appear with options to edit and format the text.

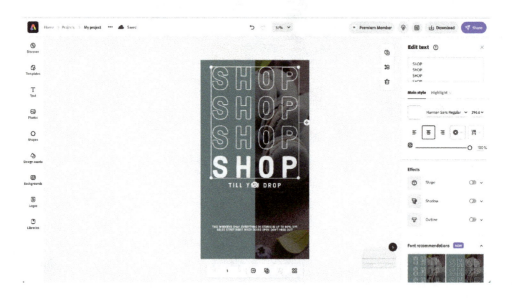

Figure 3.11 – Editing and formatting text options

15. To edit the text, simply type your text in the `Edit text` field. To change the font, click on the font arrow and select your preferred font. You also have the flexibility to change the font size, alignment, font spacing, and opacity. Additionally, you can also add effects, including shape, shadow, and outline.

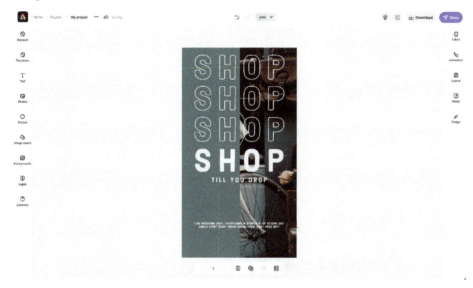

Figure 3.12 – Repurposed template to create a customized Instagram story

With Adobe Express, you can add additional elements, such as shapes, which are infographics and illustration elements. Simply click on **Shapes** on the left-hand pane and type in a keyword. Adobe Express has thousands of illustrations you can use. For this example, I typed in the word `macaroon`.

Figure 3.13 – Add infographics and illustrations to your social media graphic by browsing through shapes

16. Simply click on the shape you want to add to your project. Adobe Express will add the shape, and the **Edit shape** window will pop out on the right. Here, you can change the color, blend mode, opacity, and orientation of the shape.

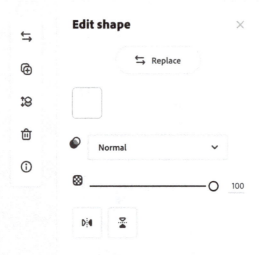

Figure 3.14 – Options to edit the shape properties

17. Change the shape properties as desired. I changed the color to white.

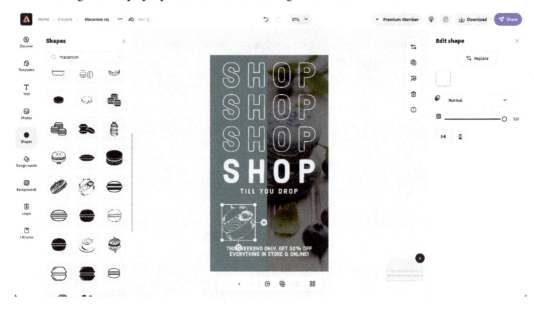

Figure 3.15 – An illustration of a macaroon has been added

18. With Adobe Express, you can also add **Design assets** to your project. You can filter the categories to one of the following: **Trending**, **Effect groups**, **Graphic groups**, **Illustrations**, **Brushes**, **Frames**, **Elements**, **Textures**, and **Overlays**.

Figure 3.16 – Adobe Express has hundreds of design assets you can use

19. For this example, I navigated to **Textures** then **Dots**, and selected one of the dot textures.

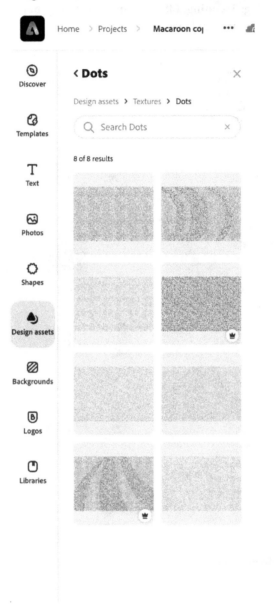

Figure 3.17 – Design assets – dot textures are one of the options available

20. Once you click on your selected design asset, Adobe Express will add this to your project. Then, on the right-hand pane, the image properties will appear. Here, you can resize the asset, change the blend mode, opacity, and orientation, and crop and shape the image as desired.

Additionally, you can also add filters, enhancements, and blur effects. These are powered by Adobe Photoshop.

Figure 3.18 – Design asset properties

21. To download your project, click on the **Download** button in the top right. Adobe Express gives you three file type options to choose from: **PNG**, **JPG**, or **PDF**. Select your preferred format, then click on **Start download**.

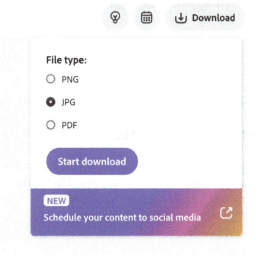

Figure 3.19 – Download options

22. Adobe Express will download a high-resolution JPG (or whichever format you picked). You can now use this beautiful graphic for your Instagram story!

Figure 3.20 – Final artwork created with an Instagram story template

In this section, you created a social media post by starting from a template in Adobe Express. With just a few clicks, you can customize the template and create your own social media post, such as an Instagram story, within minutes.

In the next section, we will learn how to create a logo by starting with a template in Adobe Express.

Creating a logo from a template

In this section, you will learn how to create a logo from a template. In the previous chapter (*Chapter 2*), you learned how to use the logo builder when creating a brand in Express. However, in this chapter and section, we will be customizing a logo template.

Following on from the previous section, you will now follow these steps to learn how to create a logo from a template:

1. Navigate to `https://express.adobe.com/` in your browser.

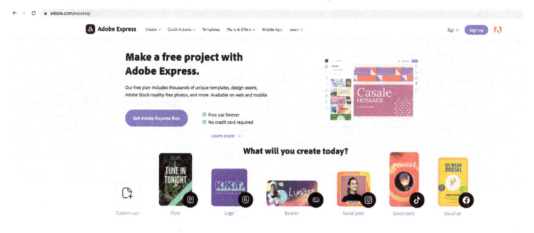

Figure 3.21 – Home page of Adobe Express on the browser

2. Click the + icon, which is located at the top left of the page.

Figure 3.22 – Click on the + icon

3. When you click on the + icon, you will be presented with options to either create new content or get started with quick actions.

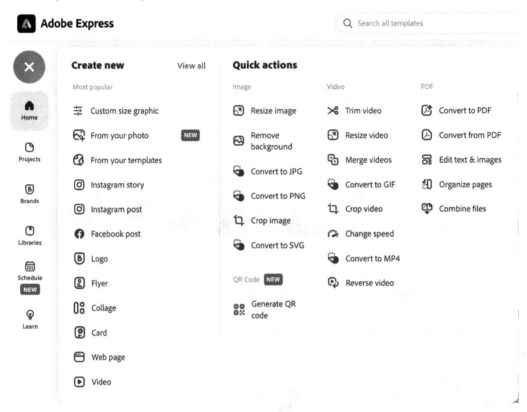

Figure 3.23 – The Create new and Quick actions options in Adobe Express

4.   Navigate to **Logo** on the **Create new** tab.

Figure 3.24 – Navigate to Logo under the Create new option

5.   Express will create a new project. You will notice that the left-hand **Templates** pane has expanded. Navigate to the arrow pointing right to expand the logo window, indicated by **See More**.

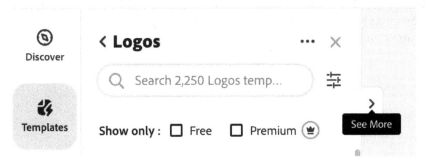

Figure 3.25 – On the left-hand pane of your project, click on the
right arrow to expand the Templates window

6.   When you click on the **See More** arrow, Express will expand the window.

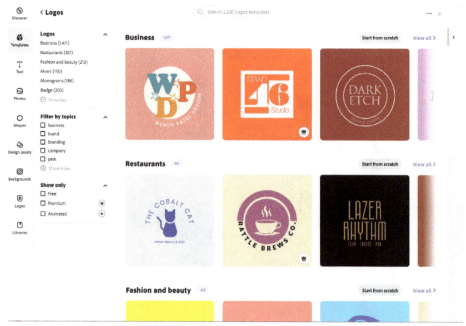

Figure 3.26 – Logo templates displayed in the project window

7.   As you can see, Adobe Express has 2,250 logo templates (at the time this book was written). You can further refine your search by filtering the logos via the following options: **Business**, **Restaurants**, **Fashion and beauty**, **Music**, **Monograms**, and **Badge**. Alternatively, you can filter by topic: **business**, **brand**, **branding**, **company**, **pink**, **identity**, **circle**, **typography**, **illustration**, and **design**.

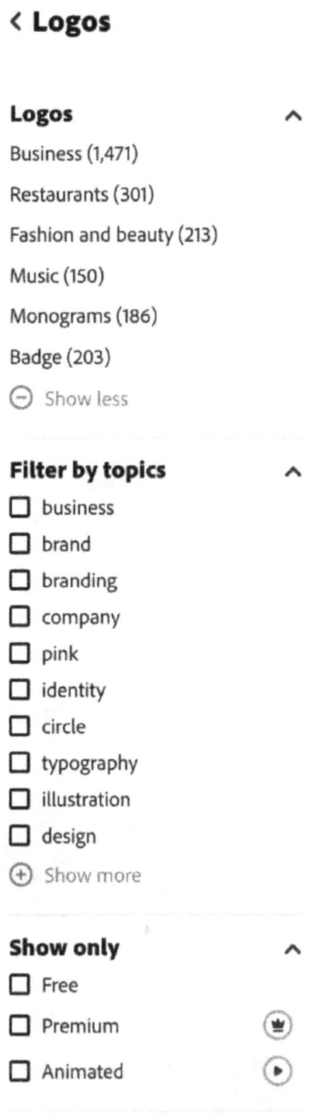

< Logos

Logos   ︿

Business (1,471)

Restaurants (301)

Fashion and beauty (213)

Music (150)

Monograms (186)

Badge (203)

⊖ Show less

Filter by topics   ︿

☐ business

☐ brand

☐ branding

☐ company

☐ pink

☐ identity

☐ circle

☐ typography

☐ illustration

☐ design

⊕ Show more

Show only   ︿

☐ Free

☐ Premium   👑

☐ Animated   ▶

Figure 3.27 – To refine your search, you can filter the templates via a category or topics

8. For this example, I will filter the logo templates to display restaurant logos by clicking on **Restaurants**.

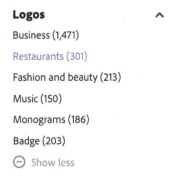

Figure 3.28 – View restaurant logo templates

9. When you click on **Restaurants**, Adobe Express will display all the restaurant logo templates available for use.

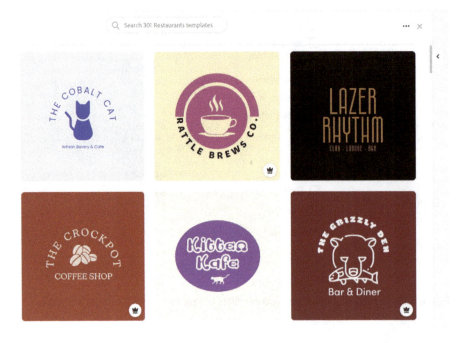

Figure 3.29 – Restaurant logo templates

10. Scroll down to view all the available restaurant templates. When you find the one you want to use, simply click on it.

11. Once you select the template, Adobe Express will display the template on the project canvas.

Figure 3.30 – Logo template displayed on the project canvas

12. To edit the text in the logo, click on the existing text on the template. This will expand the right-hand **Edit text** pane.

Figure 3.31 – To customize the text on the template, click on the existing text and this will open the Edit text window on the right-hand pane

13. Simply overwrite the existing text in the text field to replace the text. For this example, the cafe will be called Matcha Matcha. Click on the colored square to change the color of the text. Here, you can also change the font, font size, font alignment, opacity, and much more.

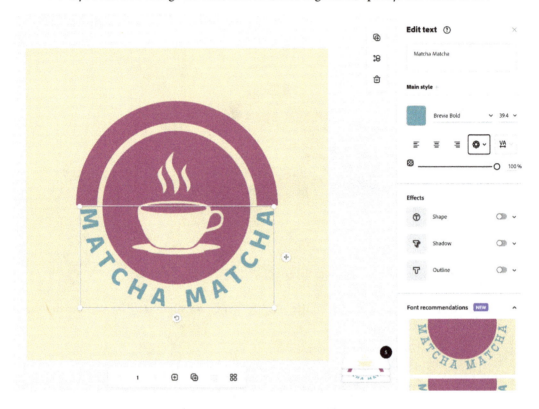

Figure 3.32 – Notice how I have changed the wording and color of the text

14. Simply click on any other element to customize the artwork. I clicked on the semi-circle. This opened up the **Edit shapes** properties in the right-hand pane.

Figure 3.33 – Click on the semi-circle to edit the shape

15. I changed the color of the shape by clicking on the pink rectangle to open the color palette.

Figure 3.34 – Select a color from the color palette to replace the existing colors on the logo template

16. Next, click on the next shape, which is the coffee mug inside the circle. The shape properties will appear in the right-hand pane.

Figure 3.35 – Click on the shape to display the shape properties

17. Click on the colored rectangle to pick your desired color for the shape.

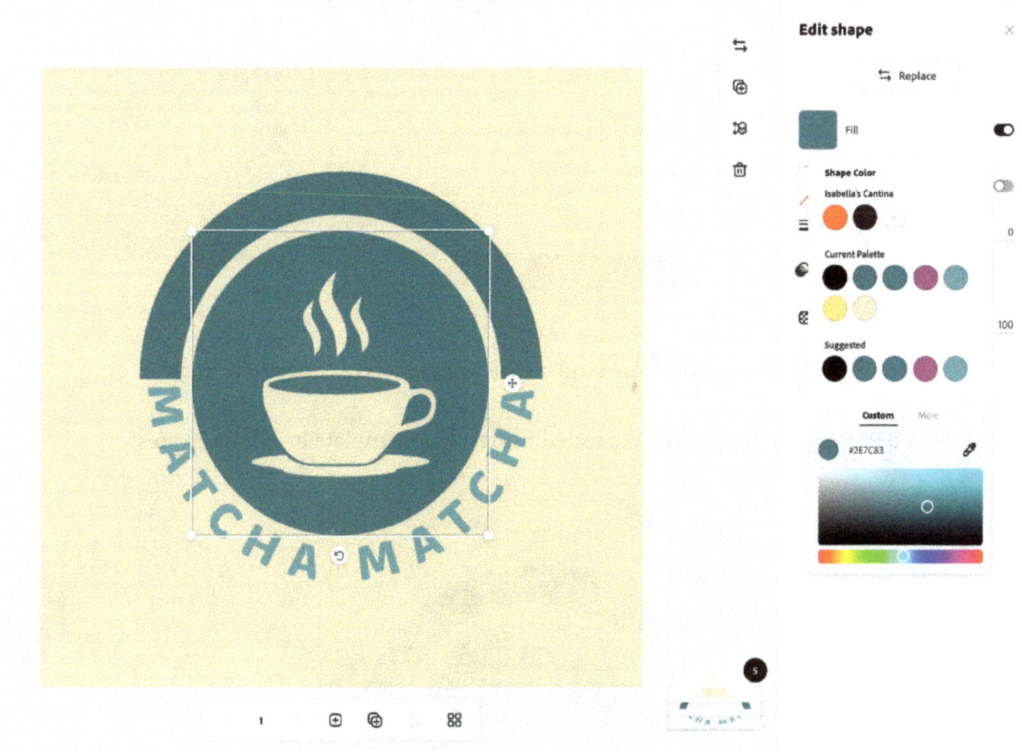

Figure 3.36 – Use the color chart to pick a color

18. That's it! Once you're happy with your customization, navigate to the top and click on the **Download** button. Adobe Express allows you to download the following formats: **PNG**, **PNG (transparent background)**, **JPG**, and **PDF**. I selected **JPG** and then clicked on the **Start download** button.

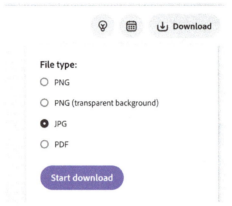

Figure 3.37 – Click on the Download button to download a copy of your logo

You now have a customized logo that you created using a logo template in Adobe Express.

Figure 3.38 – Logo that we created using a prebuilt logo template

In this section, we created a logo by starting from a template. In Express, you can choose from thousands of professionally designed logos, which you can then customize to create your own logo.

In the next section, we will explore how we can create a brochure from a template.

How to create a brochure from a template

In this section, you will learn how to create a brochure from a template. To get started, follow these steps:

1. Navigate to `https://express.adobe.com/`.

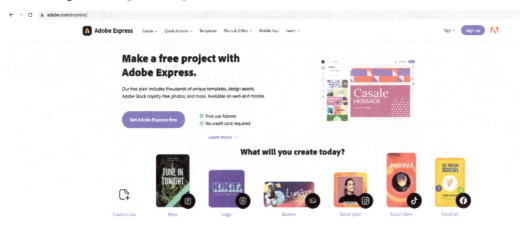

Figure 3.39 – Home page of Adobe Express on the browser

2. Click on the + icon located in the top left of the page.

Figure 3.40 – Click on the + icon

3. When you click on the + icon, you will be presented with options to either create new content or get started with quick actions.

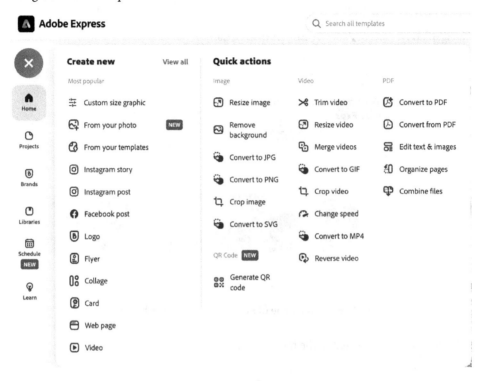

Figure 3.41 – The Create new and Quick actions options in Adobe Express

4. Navigate to the **View All** option to view all the templates.

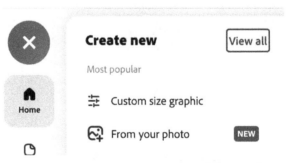

Figure 3.42 – Click on View all to display all the templates

5. Adobe Express will display all the templates and categories. Navigate to **Brochures** to view all the templates in this category.

All templates

All templates

Jump to category ⌃

Flyers (2,455)

Instagram posts (10,480)

Instagram stories (6,664)

Logos (2,250)

YouTube thumbnails (1,142)

Collages (2,222)

Facebook posts (2,261)

Facebook covers (1,061)

Cards (3,983)

Invitations (2,591)

Business cards (913)

Menus (930)

Brochures (608)

Resumes (690)

Posters (3,250)

Wallpapers (679)

Presentation graphics (400)

Album covers (793)

Book covers (794)

Worksheets (1,669)

⊖ Show less

Figure 3.43 – All the template categories available in Adobe Express

6. When you click on **Brochures**, Adobe Express will display all the templates available in the brochure category. You can further refine your search by navigating to the following: **Sales**, **Photography**, **Travel**, or **Business**. Alternatively, you can also filter by topic: **business, brochure, pink, photo, travel, information, school, advertisement, event**, and so on.

Figure 3.44 – Browse through hundreds of brochure templates in Adobe Express

7. For this example, I selected **Business** as the category. When I click on **Business**, Adobe Express will display all the business templates available for use.

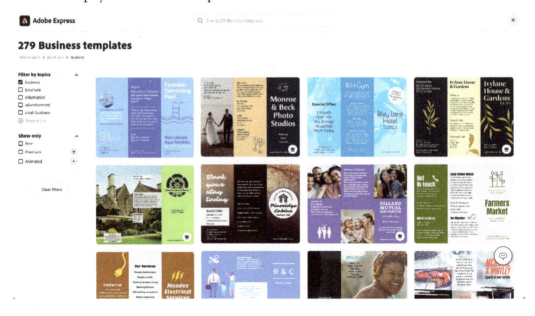

Figure 3.45 – Presentation graphics templates

8. Simply click on the template you wish to customize. When you click on a template, Adobe Express will open a new project with your selected template displayed on your project canvas.

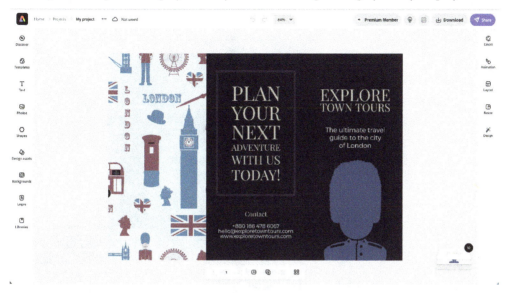

Figure 3.46 – Once you select the template, Adobe Express will display this template on the project canvas

9. Like the previous section for the logo template, all you need to do now is simply click on any of the text or shapes on the template to customize it. I clicked on the text, and this opens up the text properties in the right-hand pane.

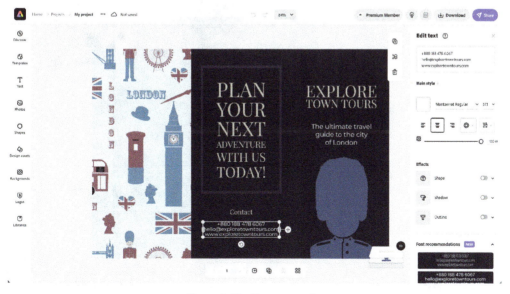

Figure 3.47 – Text properties

10. Click on any of the text to customize it.

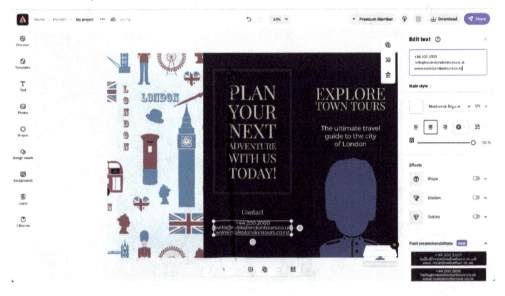

Figure 3.48 – Replace the text with your own text

11. To change the color of the shapes, simply click on it and the shape properties will be displayed in the right-hand pane.

Figure 3.49 – To edit the shape, simply click it to display the shape properties on the right-hand pane

12. Overwrite all the text and replace it with your own. Likewise, change the shapes or colors to customize the template to align it with your branding.

13. Next, click on the **Download** button. Adobe Express will display the following file formats for you to download: **PNG**, **JPG**, and **PDF**. I selected **PDF** and clicked on the **Start download** button.

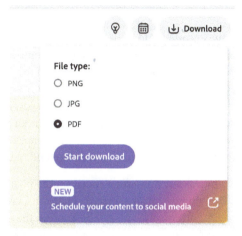

Figure 3.50 – Click on the Download button to download the file

That's it! With just a few simple clicks, you have created a beautiful brochure.

Figure 3.51 – Brochure created from a template

In this section, we learned how to create a brochure by starting from a template. Express has thousands of professionally designed templates, and with a few clicks, you can customize the templates to make them your own.

Summary

In this chapter, you have gained the knowledge to utilize the vast selection of professionally designed templates provided in Adobe Express to create an Instagram story, logo, and brochure. With Adobe Express templates at your disposal, it has never been easier to create assets for numerous outputs in a matter of minutes. You now have the freedom to create graphics in any size, resolution, and color of your preference, whether it's for personal projects or organizational needs. By leveraging the extensive collection of templates, illustrations, and color schemes in Express, you can mix and match elements to create something original. By starting with a professionally designed template, you gain a valuable head-start in the creative process, allowing you to customize any design and make your content stand out.

In this day and age, the need to pump out numerous social media posts daily is ever increasing. Additionally, with the ever-growing number of social media platforms, creators are having to create more content at an increasing speed. Fortunately, with Adobe Express, the ability to meet this demand could not be any easier. Creators don't even need to use their laptops to create content – all they need is their cellphone device to access the Adobe Express app to start creating. Furthermore, users have the convenience of directly publishing their Express creations to various social media platforms through the seamless integrations offered by the platform, including Instagram, Facebook, LinkedIn, Twitter, and Pinterest.

In the next chapter, we will explore how to level up your social media post, including ways to format text, add design assets, and change the background.

4

Level Up Your Social Media Posts with Adobe Express

In this chapter, you will gain comprehensive knowledge about modifying social media posts. By mastering techniques such as resizing posts, changing the character fonts, adding images and icons, and reformatting text, you can be assured that you will level up your social media content.

You will learn how to use the libraries of professional images licensed from Adobe Stock, icons, typography styles, and design assets to enhance your social media graphics. By honing these skills, you will elevate your social media graphics and enrich your visual communication.

At the core of graphic design lies simplicity and the ability to convey a concept through visual and textual content. By learning how to effectively format the text and images in your Express projects, you will unlock infinite possibilities of what you can achieve with your creativity.

We will cover the following topics in this chapter:

- How to change the background
- How to format text
- How to add design assets

By the end of this chapter, you will be able to create visually striking and professional designs to use on your social media platforms. You will gain proficiency in utilizing the wide array of assets at your disposal, such as illustrations and backgrounds, to enhance and elevate your design aesthetics.

How to change the background

In this section, you will learn how to change the background elements in your social media project. To get started, follow these steps:

1. Select a template from Adobe Express. Refer to the previous chapter to see how to open a template in Express. I have selected a book cover template, which I will repurpose to create a social media post.

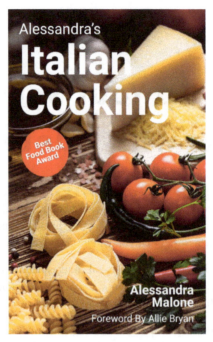

Figure 4.1 – A book cover template selected in Adobe Express

2. Once you have selected a template, rename your project. Navigate to the top-left corner of the page and click on **My project**. You will then be able to type in a new project name in the field.

Figure 4.2 – By default, your project name will be named My project

3. I have called my project Cookbook Graphic. Hit *Enter* once you have renamed your project.

Figure 4.3 – Type in the new project name (for this example, I named my project Cookbook Graphic)

4. To change the background image, simply click on the project's image. You will see a gray bounding box, with white circles on each corner. This indicates that you have selected the background image. Additionally, you will notice on the right, the contextual menu with the image formatting options will appear.

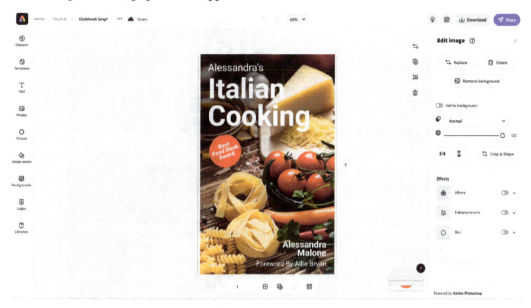

Figure 4.4 – Image formatting properties

5. When you have selected an image, Adobe Express gives you three options. You can replace the image, delete the image, or remove the background image altogether.

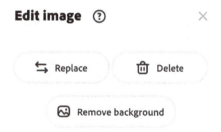

Figure 4.5 – Replace, Delete, and Remove background

6. Click on the **Replace** button.

7.   When you click the **Replace** button, Adobe Express will open the **Photos** pane on the left-hand side.

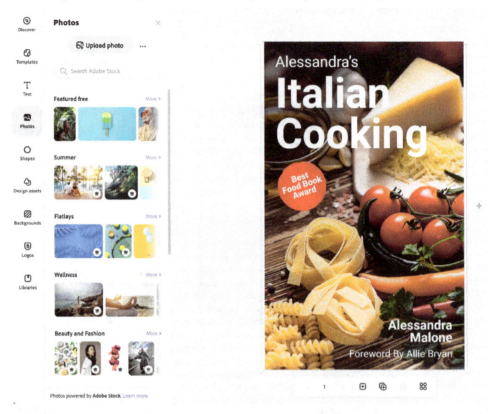

Figure 4.6 – Options to change the background image

8.   If you want to upload your own image, simply click on **Upload photo**. This will replace the existing image with your own image.

9.   Alternatively, you can click on the ellipsis. This will provide you with options to upload an image from your own Adobe Stock photos, photos from Adobe Lightroom, or an image from a third-party depository, including Dropbox, Google Drive, and Google Photos.

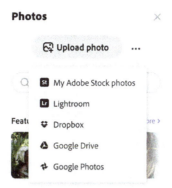

Figure 4.7 – Alternative options for uploading photos

10. Alternatively, if you don't have your own photos, you can browse through thousands of high-quality photos from Adobe Stock. Simply type in a keyword to browse their repository of images. I typed in the word `pasta` for this example.

Figure 4.8 – Browse through thousands of images from Adobe Stock

11. Simply click on the image you wish to use in your project. Adobe Express will automatically replace the existing background image. If you have a premium subscription, you have access to all the images. However, with the free subscription, you will have limited access to images (the crown icon indicates you need a premium subscription to access these images).

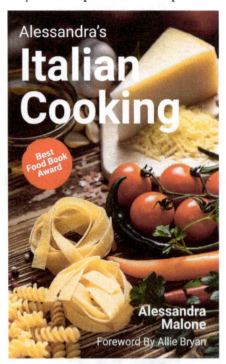

Figure 4.9 – New image I have selected from Adobe Stock to replace the existing image

12. With the image selected, you will notice the contextual menu to the right of the image. The options here allow you to perform the following actions: replace the image, duplicate the image, change the layer order, delete the image, and find the source information for the image.

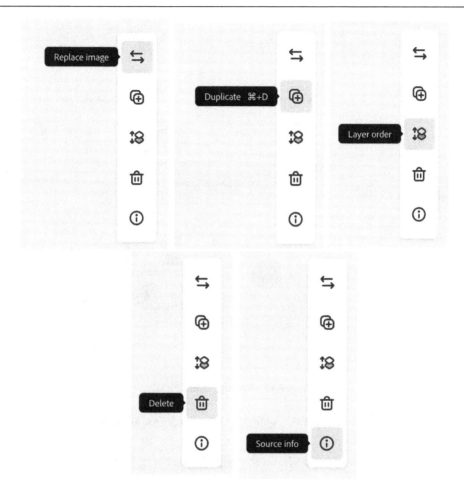

Figure 4.10 – Image options, including replace image, duplicate, and layer order

13. With the image still selected, you will also notice a flyout menu at the bottom-right corner of the image. Here, you can click on the layer (rectangle icon) once to reveal all the layers in your project. To reorder the elements, simply click and drag to reorder. For example, if you want the text on top of the image, you have to click and drag the text layer and stack it on top of the image layer.

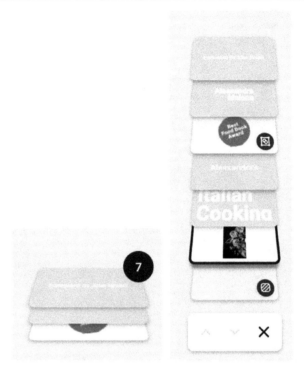

Figure 4.11 – Reorder layers

14. To fix the image to the background, toggle the **Add to background** button.

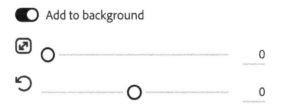

Figure 4.12 – Toggle to either fix the image to the background or float the image

15. When you toggle the **Add to background** button, you will fix the image to the background. You will then see two sliders underneath this option. The first slider allows you to resize the image, and the second slider allows you to rotate the image.

Figure 4.13 – Sliders for adjusting the size and rotation of the image

16. However, if you toggle the **Add to background** option off, you will allow the image to float and you will have the flexibility to move the image around freely with your cursor.

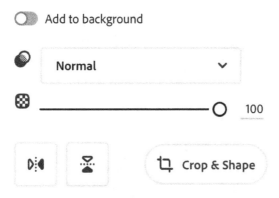

Figure 4.14 – Add to background option toggled off

17. When you have the **Add to background** button toggled off, you will be able to change the blend mode of the image to **Normal**, **Multiply**, or **Screen**.

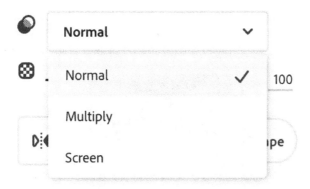

Figure 4.15 – Normal, Multiply, and Screen blend modes

18. To demonstrate the effects of the different blend modes, I have added an extra image of cheese on top of the background image to illustrate the differences between them:

- **Normal** blend mode leaves your image in its original state. You will notice the image of the cheese is placed directly on the background image.

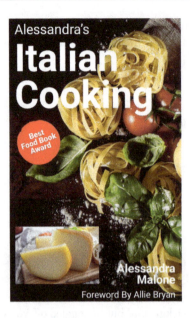

Figure 4.16 – Normal blend mode

- The **Multiply** blend mode multiplies the colors of the blending layer and the base layers, which results in a darker color. As you can see in this example, you can hardly see the image of the cheese because it is sitting on a dark area of the background image:

Figure 4.17 – Multiply blend mode

- I enlarged the image and moved it to the top half of the background image. Here, you can see the cheese image blends with the background image:

Figure 4.18 – Multiply blend mode with the cheese image in the top half of the original image

- **Screen** blend mode inverts both layers, multiplies them, and then inverts the result.

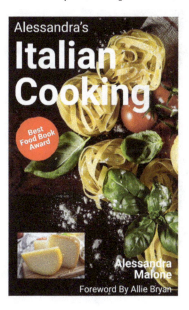

Figure 4.19 – Screen blend mode

In this section, we explored how to change the background image of any project. In Express, there are many ways of importing an image, including uploading your own photos, choosing an image from Adobe Stock, and uploading an image from third-party integrations, such as Google Drive. We looked at the different ways of fixing and floating images on the canvas. This flexibility allows us to fix an image to the background or place an image as a standalone image in a project. Finally, we also looked at the powerful Photoshop integration in Express, where we can manipulate the blend modes for an image.

In the next section, we will explore how to format the text in your project. Just like with images, Express gives you the flexibility to format text in many different ways.

How to format text

In this section, you will learn how to format the text in your social media post project. In the previous chapter, we introduced some of the formatting options. However, in this chapter, we will take a more in-depth look at all the formatting options.

For this section, I will use the cookbook project that I used in the previous section.

To get started, follow these steps:

1.   Click on the text **Italian Cooking**. This will open the text formatting properties in the contextual menu on the sidebar.

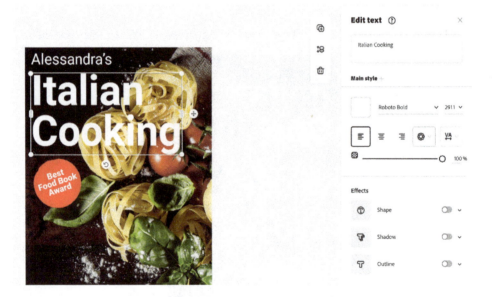

Figure 4.20 – When you select the text, the text editing properties will open in the right-hand pane

2. In the text field, you can replace the text by overwriting it. This will update the text on the canvas. I left the text as `Italian Cooking`.

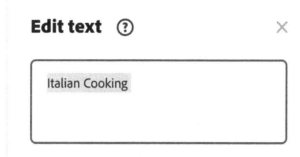

Figure 4.21 – Replace the existing text in the text field

3. To change the text color, select the square below **Main style**. In this example, the square is white (which is the color of the text on the canvas).

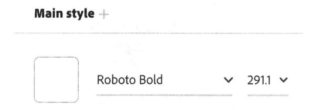

Figure 4.22 – Change the font color by selecting the square

4. When you click on the square, the text color window will open. You will see your brand's colors at the top (if you have set up your brand in Express). In the example here, you can see the fictitious brand I created called **Isabella's Cantina**. Next, you will see the colors displayed in the project under **Current Palette**. Beneath that, you will see **Suggested** colors.

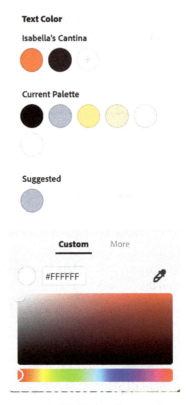

Figure 4.23 – Text color options

5.   At the very bottom, you will notice a color picker, where you can select a color with your cursor. This is under the **Custom** option. Additionally, you can use the eyedropper tool to pick a color from your project. Finally, you can also insert the color hex code if you have a specific color that you want to use in your project.

Figure 4.24 – Custom color picker

6. You can click on **More** to display colors in individual circles for more precise selection.

Figure 4.25 – More color options

7. To change the font, click on the drop-down arrow to display the hundreds of fonts available from Adobe Fonts.

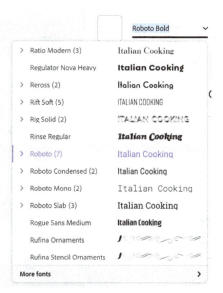

Figure 4.26 – Choose from hundreds of fonts available from Adobe Font

8.   To change the text size, click on the drop-down box with the font size value. You can also type in a font size.

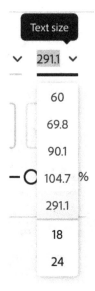

Figure 4.27 – Font size

9.   To change the text alignment, simply click on the paragraph icons to align left, center, or right.

Figure 4.28 – Text alignment options

10. For more text alignment options, click on the donut/circle icon to display the curved, grid, and magic alignment options.

Figure 4.29 – Curved, grid, and magic alignment options

11. Click on this donut icon to display the various alignment options. The **Curved** alignment options include circle, semi-circle, and inverted semi-circle. The **Grid** alignment options include left-aligned grid, center-aligned grid, and right-aligned grid. The **Magic** alignment options include capitalize and fit, rotate, and drop cap.

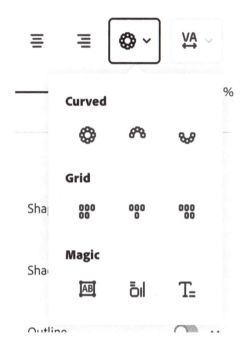

Figure 4.30 - Curved, grid, and magic alignment options

12. Here are some examples of what the **Magic** alignments look like:

Figure 4.31 – Circle alignment, rotate alignment, and drop cap alignment

13. To change the line spacing, navigate to the **Spacing** option, which is disignated by the icon with the arrow and **VA**.

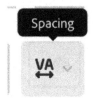

Figure 4.32 – Text spacing option

14. When you click on the spacing drop-down menu, you will have the ability to adjust the line and letter spacing of the text.

Figure 4.33 – Line and letter spacing

15. Next, let's explore the text effects options, which include the following: **Shape**, **Shadow**, and **Outline**.

Figure 4.34 – Effects include Shape, Shadow, and Outline

16. To add a shape to the background of your text, toggle the **Shape** slider.

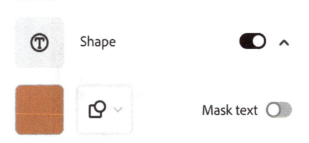

Figure 4.35 – Toggle the slider to view the Shape alignment options

17. Click on the circle and square icon to display all the shape alignment options. Explore how each shape looks by simply clicking on it. This will be reflected immediately on the canvas. You can also change the size and opacity of the shape.

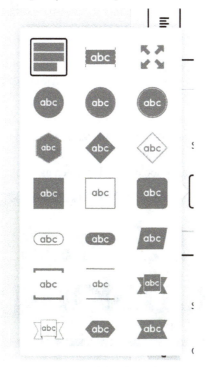

Figure 4.36 – Text shape options

18. For example, the following screenshot displays the text with a rectangular background.

Figure 4.37 – Rectangle shape

19. You can also mask the text by toggling the **Mask text** option.

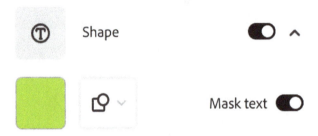

Figure 4.38 – Toggle the Mask text option

20. Here is an example of what the text looks like when you mask the text on a rectangular background.

Figure 4.39 – Masked text example

21. To add shadow to your text, simply toggle the **Shadow** option to display the options. Here, you can change the color, angle, and distance of the shadow.

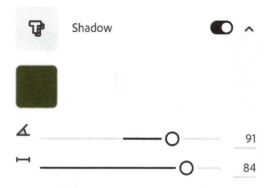

Figure 4.40 – Text shadow options

22. This is an example of text with a shadow applied to it:

Figure 4.41 – Text shadow

23. To change the outline of the text, toggle on the **Outline** option. You can pick a color and adjust the thickness of the outline by using the slider.

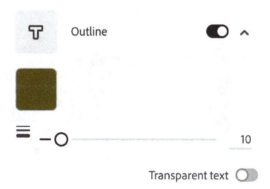

Figure 4.42 – Text Outline options

24. This is what the text looks like with an outline applied:

Figure 4.43 – Outline applied to the text

25. Toggle the **Transparent text** slider to remove the fill from your text.

Figure 4.44 – Toggle the Transparent text option

26. This is an example of what the final text looks like with the **Transparent text** option toggled on.

Figure 4.45 – Transparent text applied to the example

27. At the very bottom, you can see some font recommendations. Simply click on any of these options to apply it to your text.

Figure 4.46 – Font recommendations

In this section, you learned how to format text in Express. We explored the many text formatting functions in Express that make it easy for you to create visually appealing content. Text can overwhelm the viewer with information. However, with Express, there are many formatting options that guide you and enable you to create a beautiful aesthetic for your audience.

In the next section, we will explore how to utilize the professionally designed design assets provided in Express.

Adding design assets

In this section, you will learn how to add design assets from Adobe Express to make your content stand out from the crowd.

To get started, follow these steps:

1. Navigate to the left-hand pane and select **Design assets**. Adobe Express will open the **Design assets** window.

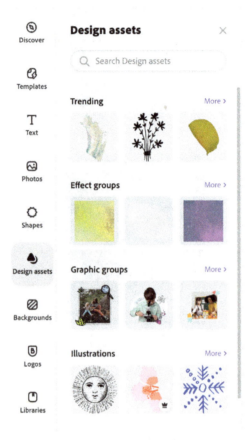

Figure 4.47 – Design assets

2. To display assets from any of the categories, simply click on the **More** button. I clicked on **Illustrations**.

3. Once you click on one of the categories, such as illustrations, Adobe Express will display more filters to further refine your search. For illustrations, for example, you can filter via the following filters: **Nature**, **Animals**, **People and Beings**, **Hearts**, **Stars**, **Sports and Activities**, **Food and Drink**, **Objects**, **Transportation**, **Seasonal**, and **Travel & Adventure**.

Figure 4.48 – Browse through hundreds of Illustrations in Adobe Express

4.   You can also use the search box and type in a keyword to search for an illustration.

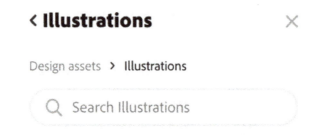

Figure 4.49 – Search by a keyword

5.   For this example, I navigated to the **Savory And Salty Food** category and picked a croissant illustration. Simply click on the illustration you want to add to your project.

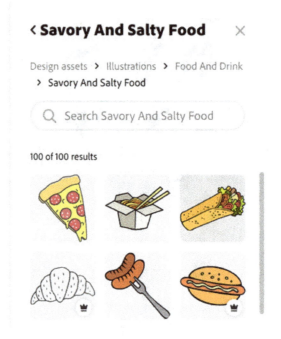

Figure 4.50 – Savory and Salty Food illustrations

6. The illustration I picked has been added to my project:

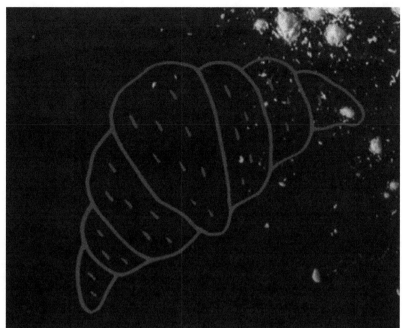

Figure 4.51 – Croissant illustration

In this section, we explored how easy it is to add professionally designed design assets to your project. The design assets you can find in Express include a wide array of graphics from illustrations to frames, textures, and overlays. By adding these elements to your design, you can elevate any artwork to the level of professional graphic design. Or, if you are a graphic designer, you can also use design assets to supplement your original creations.

Summary

In this chapter, we explored ways to customize a social media post by changing the background image, formatting the text, and adding design elements to your project. With access to these versatile formatting tools, you open up endless possibilities to create standout content that captivates your audience. Starting from a template and taking advantage of the flexibility to customize all these elements, you can create content that truly stands out from the rest.

In the next chapter, we will explore how to convert your projects into simple animations. You will gain the expertise to export your animations as GIFs or MP4 files, allowing you to seamlessly integrate them into your social media posts.

5

Animating Text and Images with Adobe Express

In this chapter, you will learn how to repurpose any content created in Adobe Express by converting it into a simple animation. These animations are five seconds long, which is great for social media posts, such as Instagram stories, posts, reels, LinkedIn posts, TikTok videos, Facebook ads, and much more.

You will first learn how to create text animations by selecting from a variety of inbuilt animation presets, which include the following effects: *typewriter, dynamic, flicker, color shuffle, fade, slide,* and *grow*. Additionally, you will also learn how to add animation to images by choosing from a variety of inbuilt animation presets, which include the following effects: *zoom, pan, grey, blur, color,* and *fade*.

Finally, we will also learn how to export the animations as MP4 files, enabling them to be effortlessly shared on various social media platforms.

We will cover the following topics in this chapter:

- Creating animations for text and images
- How to export animations

By the end of this chapter, you will be able to create stunning professional-looking animated GIFs and animated MP4 files. Using Adobe Express, you can create sophisticated animations without any heavy lifting or knowledge of animation. This empowers you to create standout content that captivates your audience, compelling them to pause and engage with your visually appealing animated creations.

Creating animations for text and images

In this section, we will explore the process of adding dynamic animations to your text and image content. With Adobe Express, you can easily add text and photo animations with just a few clicks, and generate MP4 video files.

How to create text animations

In this section, you will be introduced to adding motion to your text, which can bring new life to your projects. With just a few clicks, you will be able to add dynamic movement to your text, making it more engaging to your audience. This technique is super easy to master, so let's get started.

To fully experience the animations shown in this chapter, either scan the following QR code or visit `https://bit.ly/adobeexpressanimations`, where you can view all the animations in their entirety.

Figure 5.1 – QR code for the animations links

To get started with adding animations to our text, follow these steps in your browser:

1. Open any of your existing projects by navigating to the **Projects** tab in the left-hand pane. Alternatively, refer to *Chapter 3* to see how to create a graphic from a template in Adobe Express:

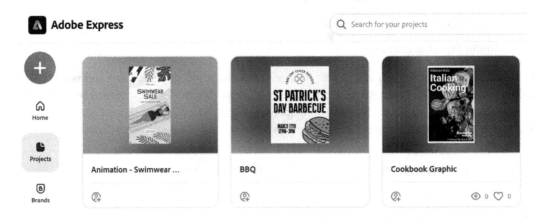

Figure 5.2 – Projects will display all your projects on the home page of Adobe Express

2. Hover your cursor over the project you want to edit, then click on **Edit project** to open the project:

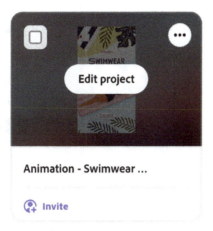

Figure 5.3 – To open a project, click on the Edit project button

Adobe Express will open the project:

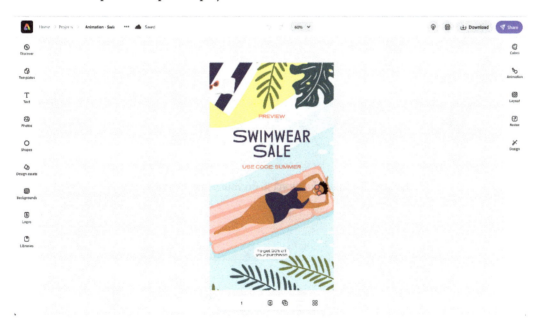

Figure 5.4 – Project page

3.   In the right-hand pane, navigate to the **Animation** tab:

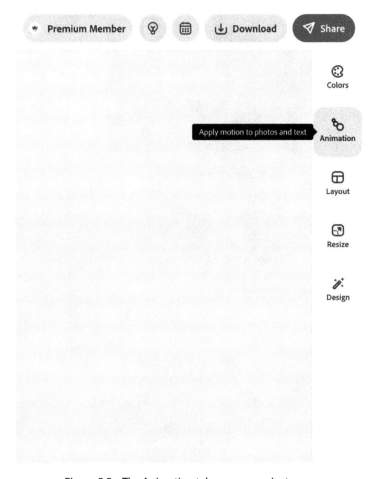

Figure 5.5 – The Animation tab on your project page

Click on **Animation** to open the **Animation** properties window:

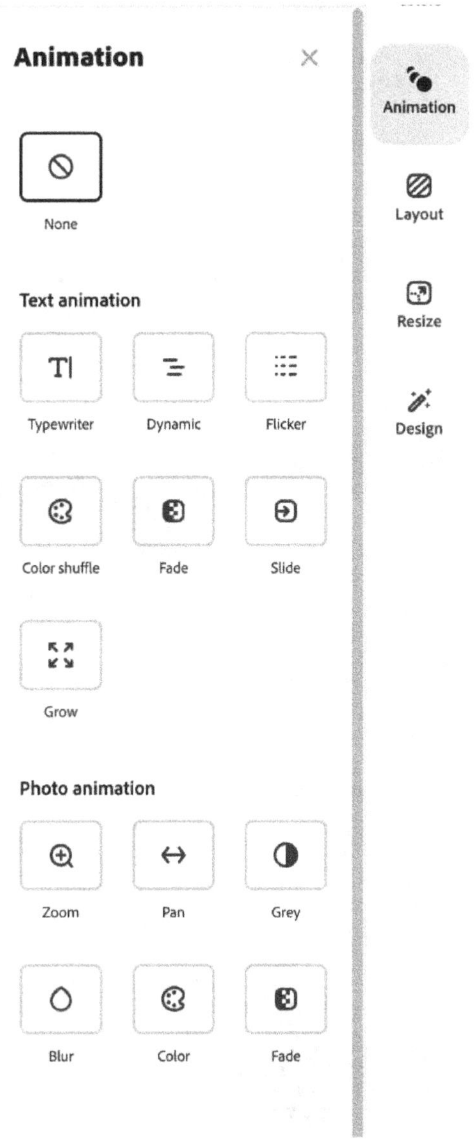

Figure 5.6 – Animation window

As you can see in *Figure 5.5*, the **Animation** options fall into two categories, **Text animation** and **Photo animation**. In this section, we will explore text animations.

4. Click on **Typewriter**. Adobe Express will automatically animate each character of the word with a *typewriter* animation. Each character will appear as if the word is being typed in real time.

Figure 5.7 – Typewriter text animation in motion (left) and final view (right)

Figure 5.8 – To view the animation, scan the QR code or click on the 01 Typewriter -
Text Animation MP4 file in this folder: https://bit.ly/adobeexpressanimations

5.	Click on **Dynamic**. Adobe Express will animate the words so that they appear one word at a time in a *dynamic* effect.

Figure 5.9 – Dynamic text animation in motion (left) and final view (right)

Figure 5.10 – To view the animation, scan the qr code or click on the '02 Dynamic - Text Animation' MP4 file in this folder: https://bit.ly/adobeexpressanimations

6.   Click on **Flicker**. Adobe Express will animate the words so they flicker like disco lights.

Figure 5.11 – Flicker text animation in motion (left) and final view (right)

Figure 5.12 – To view the animation, scan the QR code or click on the '03 Flicker - Text
Animation' MP4 file in this folder: https://bit.ly/adobeexpressanimations

7.   Click on **Color shuffle**. Adobe Express will animate the words by rotating the colors of the words.

Figure 5.13 – Color shuffle text animation in motion (left) and final view (right)

Figure 5.14 – To view the animation, scan the QR code or click on the 04 Color shuffle -
Text Animation MP4 file in this folder: https://bit.ly/adobeexpressanimations

8.   Click on **Fade**. Adobe Express will automatically animate the text so that each word fades in and appears on the screen.

Figure 5.15 – Fade text animation in motion (left) and final view (right)

Figure 5.16 – To view the animation, scan the QR code or click on the '05 Fade - Text Animation' MP4 file in this folder: https://bit.ly/adobeexpressanimations

9.   Click on **Slide**. Adobe Express will automatically animate the text so that it slides in from the left.

Figure 5.17 – Slide text animation in motion (left) and final view (right)

Figure 5.18 – To view the animation, scan on the QR code or click on the '06 Slide -
Text Animation' MP4 file in this folder: https://bit.ly/adobeexpressanimations

10. Click on **Grow**. Adobe Express will automatically animate the text so it grows in size.

Figure 5.19 – Grow text animation in motion (left) and final view (right)

Figure 5.20 – To view the animation, scan the QR code or click on the '07 Grow - Text
Animation MP4' file in this folder: https://bit.ly/adobeexpressanimations

In this section, you mastered the basics of adding motion to text, a technique that can add excitement to projects and make your project more engaging for your audience. With this fun and easy technique, you can easily elevate your projects with this dynamic addition.

How to create image animations

To view all the animations shown in this chapter, please visit this folder: `https://bit.ly/adobeexpressanimations`.

To get started with adding animation to your images, follow these steps in your browser:

1. *Steps 1-3* are the same as described in the previous section, *How to create text animations*. As mentioned in the previous section, the **Animation** options fall into two categories: text animations and photo animations. In this section, we will explore photo animations since we are animating our images now.

2. Click on **Zoom**. Adobe Express will automatically add animation to the photo and zoom in on the content.

Figure 5.21 – Zoom photo animation in motion (left) and final view (right)

Figure 5.22 – To view the animation, scan the QR code or click on the '07 Grow - Text
Animation MP4' MP4 file in this folder: https://bit.ly/adobeexpressanimations

3.   Click on **Pan**. Adobe Express will automatically pan to the content so that the image appears
to be sliding from the right.

Figure 5.23 – Pan photo animation in motion (left) and final view (right)

Figure 5.24 – To view the animation, scan the QR code or click on the '09 Pan - Photo Animation' MP4 file in this folder: https://bit.ly/adobeexpressanimations

4. Click on **Grey**. Adobe Express will automatically add an animation to the photo in which the image starts off in grayscale and transitions into full color.

Figure 5.25 – Grey photo animation in motion (left) and final view (right)

Figure 5.26 – To view the animation, scan the QR code or click on the '10 Grey - Photo
Animation' MP4 file in this folder: https://bit.ly/adobeexpressanimations

5. Click on **Blur**. Adobe Express will automatically add an animation where the image appears
    blurred to begin with, then transitions into sharp focus.

Figure 5.27 – Blur photo animation in motion (left) and final view (right)

Figure 5.28 – To view the animation, scan the QR code or click on the '11 Blur - Photo Animation' MP4 file in this folder: https://bit.ly/adobeexpressanimations

6. Click on **Color**. Adobe Express will automatically add an animation that gradually saturates the colors in the image.

Figure 5.29 – Color photo animation in motion (left) and final view (right)

Figure 5.30 – To view the animation, scan the QR code or click on the '12 Color - Photo
Animation' MP4 file in this folder: https://bit.ly/adobeexpressanimations

7. Click on **Fade**. Adobe Express will automatically add an animation where the image fades
    into view.

Figure 5.31 – Fade photo animation in motion (left) and final view (right)

Figure 5.32 – To view the animation, scan the QR code or click on the '13 Fade - Photo Animation' MP4 file in this folder: https://bit.ly/adobeexpressanimations

In this section, you learned how to add animation to your text and images. With just a few clicks, you can easily make your content dynamic by adding simple animations. With this Express feature, you can create exciting content and stand out from the crowd. In the past, creating animations required extensive skills in motion graphics and proficiency with complex tools such as Adobe After Effects. However, Express has eliminated this steep learning curve, enabling everyone to effortlessly create basic animations.

In the next section, we will explore how to export animations as MP4 files, making them simple to download or share on various social media platforms.

How to export animations

Now that we have added animations to our text and images, it is time to prepare our project so we can share it on social media. In this section, you will learn how to export your project as an MP4 file. This step will enable seamless sharing of your animations, showcasing your creativity to your targeted audience.

Follow these steps:

1. Navigate to the top-right corner to locate the **Download** button:

Figure 5.33 – The Download button

2. Click on the **Download** button. This will display the various file types you can download. We want to download the **MP4 Video (720p)** file.

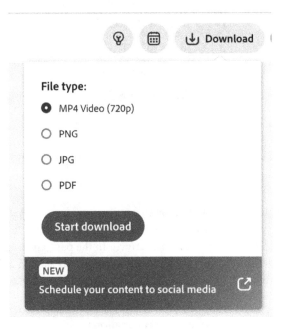

Figure 5.34 – File types you can download

3. Click on **MP4 Video (720p)**, then click on the **Start download** button.

4. Navigate to your Downloads folder to view your downloaded MP4 video file.

Figure 5.35 – View the animations by scanning the QR code or via
this link: https://bit.ly/adobeexpressanimations

Summary

In this chapter, you learned how to add animations to your text and image content. With just the click of a button, Adobe Express adds text and photo animations quickly and easily. Gone are the days of spending hours in Adobe After Effects to create these captivating yet straightforward dynamic animations. By navigating to the **Animation** tab, you can select the desired animation style to enhance your content. Adobe Express automatically generates an MP4 video file for you to download, so you can quickly upload these polished animations to your social media platforms.

In the next chapter, you will learn how to edit images using quick actions. These actions take advantage of the power of Adobe's core apps, which include Photoshop, Premiere Pro, and Acrobat, to streamline your image editing workflow.

6
Editing Images Using Quick Actions

Powered by Adobe's core apps, which include Photoshop, Premiere Pro, and Acrobat, Adobe Express now grants you access to a host of robust photo, PDF, and video editing functionalities. These powerful features are integrated within Adobe Express and can be found under the umbrella of Quick Actions.

Harnessing the power of Photoshop, functionalities such as resizing an image, removing backgrounds, and converting JPGs into PNGs are among the plethora of photo editing tools that are readily available at your fingertips.

Powered by Premiere Pro, you can explore functionalities such as video-to-GIF conversion, video cropping and resizing, and video speed adjustment; these are just some of the comprehensive video editing tools within Adobe Express.

And finally, harnessing the power of Acrobat, tasks such as converting a PDF into a JPG and vice versa are just some of the PDF editing capabilities you can utilize within Adobe Express.

We will cover the following topics in this chapter:

- How to easily remove the background from an image
- How to resize and crop an image
- How to convert PNGs into JPGs

By the end of this chapter, you will have honed the skills to employ the versatile Quick Actions tools available in Adobe Express. You will be able to remove the background from any image, transform images into your preferred dimensions, and acquire the skills necessary to optimize file sizes by converting PNG formats into JPGs, as well as enhance the file resolution by converting JPGs into PNGs.

Learning how to easily remove the background from an image

To get started, follow these steps in the browser:

1. Navigate to the Adobe Express home page, https://express.adobe.com/, in your browser:

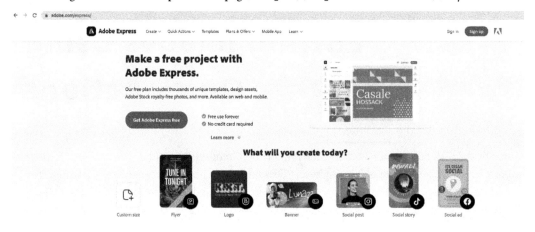

Figure 6.1 – Access Adobe Express via the browser

2. Navigate to the + icon:

Figure 6.2 – Click on the + icon to open the Quick Actions window

3. When you click on the + icon, the **Quick actions** tab opens:

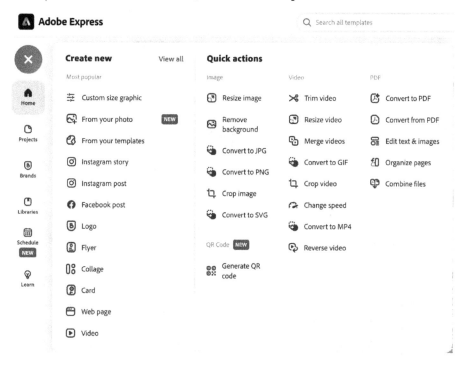

Figure 6.3 – Quick actions options

4. Under **Quick actions**, navigate to the **Remove background** option.

Figure 6.4 – The Remove background option

5. When you click on **Remove background**, the following dialog box will appear:

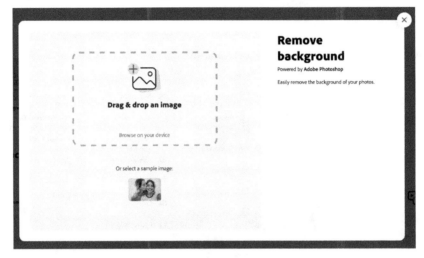

Figure 6.5 – The Remove background window

6. Simply click on **Browse on your device** to select your image, then click **Open**.

Figure 6.6 – Select an image from your device

7. Adobe Express will process this and start removing the background.

Figure 6.7 – Adobe Express will display the progress of removing
the background from the image you have selected

8. Once Adobe has finished processing this, you will see two buttons, **Customize** and **Download**:

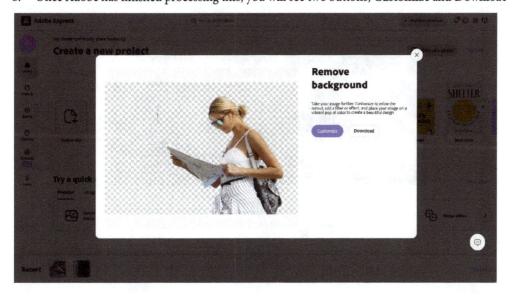

Figure 6.8 – Once Adobe has removed the background, you can either customize or download the image

9. Click on the **Download** button. Adobe Express will download a PNG file (with a transparent background).

Figure 6.9 – Copy of the image without the background

10. At the same time, Adobe will open the image in a new project. Here, you have the option to continue editing. Click on the **Keep editing** button:

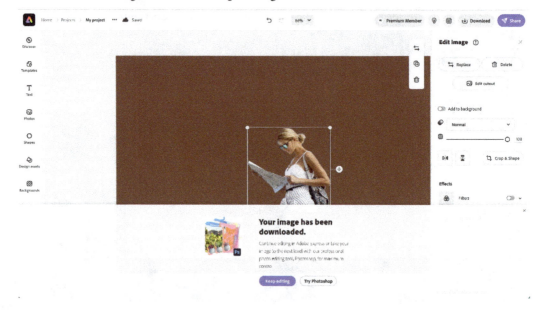

Figure 6.10 – Adobe Express will open the image in a new project window

11. Click on the background, then, in the right-hand pane, select the **Choose image** button.

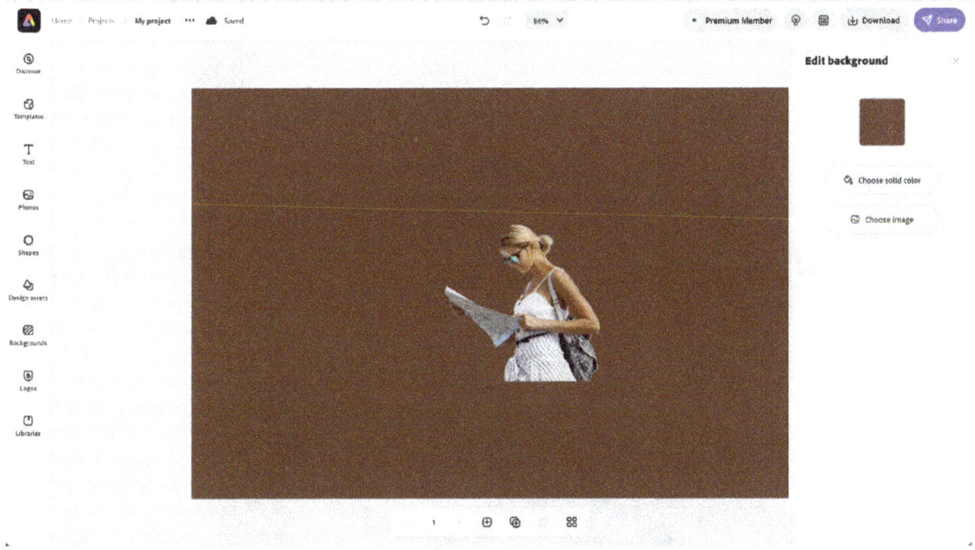

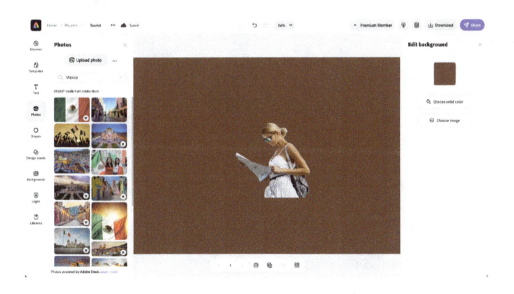

Figure 6.11 – Edit background properties

12. This will open up the **Photos** tab on the left-hand pane where you can search for images in Adobe Stock. I typed in the keyword Mexico.

Figure 6.12 – Search through millions of Adobe Stock images

13. Simply click on the thumbnail of the image of your choice and Adobe Stock will automatically add this to your project. You can click on the subject and resize and position her so she fits into the new background.

Figure 6.13 – Once you choose an image, Adobe will replace the background image

14. Click on the **Download** button, then click on **JPG** and click on the **Start download** button.

Figure 6.14 – Download options

15. You now have a high-resolution image of your new photo composite, which you can create in a matter of minutes.

Figure 6.15 – Final image of the tourist with the background from the original image
removed and replaced with a new image of a location in Mexico from Adobe Stock

In this section, we looked at how easy it is to remove the background from an image. Removing the background from an image is one of the most common image manipulation requirements, and in the past, individuals had to use powerful image editing software, such as Photoshop, to perform this task. However, with Express, with a few clicks, anyone can remove the background from an image easily.

In the next section, we will explore how to resize and crop images in Express.

How to resize and crop an image

In this section, we will explore how you can resize and crop an image using Adobe Express. To start with, let's explore how you can crop an image using Quick Actions.

To get started, follow these steps in the browser:

1.   Navigate to the Adobe Express home page, `https://express.adobe.com/`, in your browser:

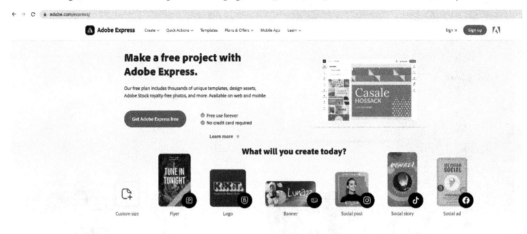

Figure 6.16 – Accessing Adobe Express via the browser

2.   Navigate to the + icon.

Figure 6.17 – Click on the + to open the Quick Actions window

3. When you click on the + icon, the **Quick actions** tab opens.

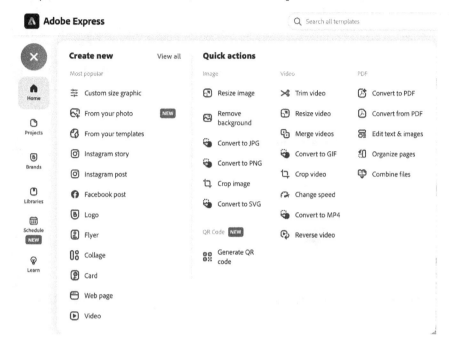

Figure 6.18 – Quick actions options

4. Under **Quick actions**, navigate to the **Resize image** option.

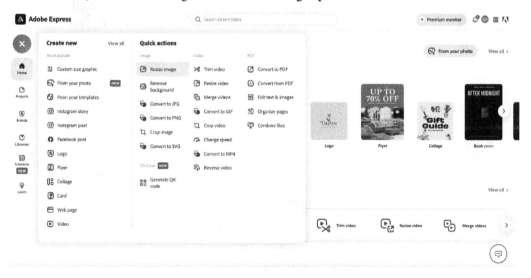

Figure 6.19 – Resize image under Quick actions

5. Adobe Express will open the **Resize image** window. You can either drag and drop an image or click on **Browse on your device**.

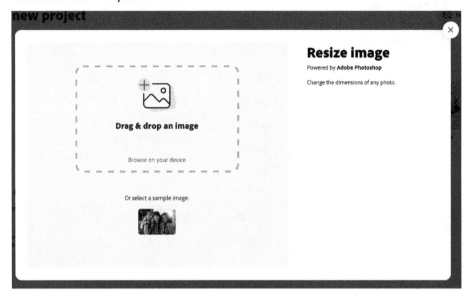

Figure 6.20 – Resize image window

6. Select an image from your hard drive, then click **Open**.

Figure 6.21 – Select an image from your hard drive, then click Open

7. Adobe Express will display a window with image resizing options.

Figure 6.22 – Resize image options

8. If you click on the **Resize for** drop-down, Adobe Express will display presets for the following social media platforms: **Instagram**, **Facebook**, **Twitter**, **YouTube**, **Pinterest**, **LinkedIn**, and **Snapchat**. Additionally, you can select **Standard** or **Custom**.

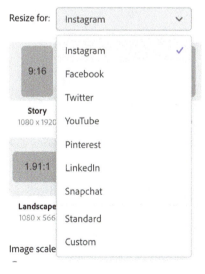

Figure 6.23 – Drop-down menu with Social Media presets

9.   When you click on any of the options, let's take **Instagram** for example, Adobe Express will display all the sizes for Instagram, which includes the following preset sizes: **Story**, **Square**, **Portrait**, and **Landscape**.

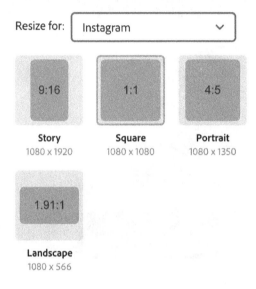

Figure 6.24 – Story, Square, Portrait, and Landscape Instagram presets

10.   When you click on **Standard**, Adobe Express will display the standard image sizes, which include **Widescreen (16:9)**, **iPhone (9:16)**, **Presentation Slide (4:3)**, **Square (1:1)**, **Landscape (3:2)**, and **Portrait (2:3)**.

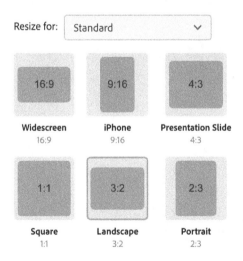

Figure 6.25 – Widescreen, iPhone, Presentation Slide, Square, Landscape, and Portrait Standard sizes

11. When you click on **Custom**, you can input your own values in pixels.

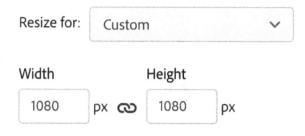

Figure 6.26 – Custom size

12. For this example, I selected **Instagram** and **Portrait**. Adobe Express will automatically adjust the image to fit the Instagram Portrait aspect ratio of 1,080 x 1,350 px.

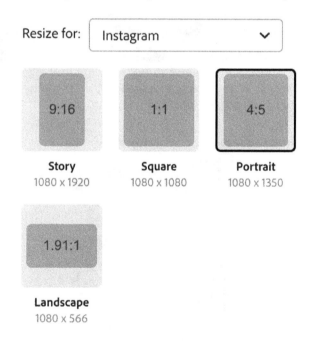

Figure 6.27 – Simply click on the size you want and Adobe Express will automatically resize the image

13. When you click on the image, your cursor changes into a hand icon, which indicates that you can click and drag to reposition the image (otherwise known as *panning*).

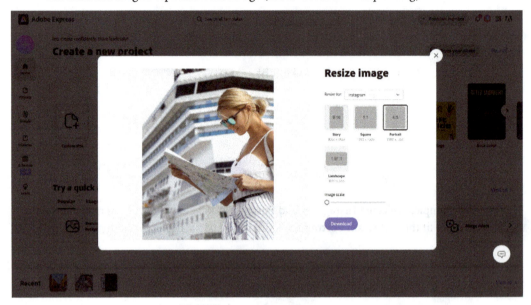

Figure 6.28 – Adobe Express has resized the image to fit an
Instagram portrait, with 1,080 x 1,350 px dimensions

14. Click on the **Download** button to download a copy of the image.

Figure 6.29 – Resized image to an Instagram portrait

Next, let's explore how to crop an image using Quick Actions:

1. Follow *steps 1-3* in the previous process.

2. Click on **Crop image** in the **Quick actions** menu.

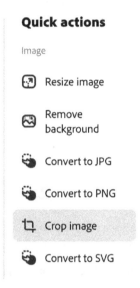

Figure 6.30 – Crop image under Quick actions

3. When you click on **Crop image** under **Quick actions**, Adobe Express will open the **Crop image** window.

Figure 6.31 – Crop image window

4. You can drag and drop an image, or you can click on **Browse on your device** to upload an image.

5. Navigate to the image you want to upload, then click **Open**.

Figure 6.32 – Uploading an image

6. Adobe Express will display the image with the crop handles, including grids. Simply pull the corners to crop the image.

Figure 6.33 – Crop image window

7. Click and drag the corners to crop your image. Adobe Express will show you a preview by dimming the background outside of your crop marks.

Figure 6.34 – Pull the corners to crop your image

8. To process the crop, simply click on the **Download** button.

9. Adobe Express will display a window to notify you when your image has been downloaded.

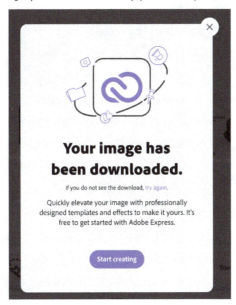

Figure 6.35 – Notification that your image has been downloaded

10. Navigate to your downloads folder to retrieve the cropped image.

Figure 6.36 – Final cropped image

In this section, we learned how to resize and crop an image in Express. Additionally, we learned that Express provides preloaded standard image dimensions, including dimensions for all the social media platforms. By incorporating these features, Express has taken the guesswork out of resizing your projects, making it super-easy for anyone to crop and resize their images.

In the next section, we will explore how to convert PNG files into JPG files.

How to convert PNG files into JPG files

To get started, follow these steps in the browser:

1. Navigate to the Adobe Express home page, `https://express.adobe.com/`, in your browser:

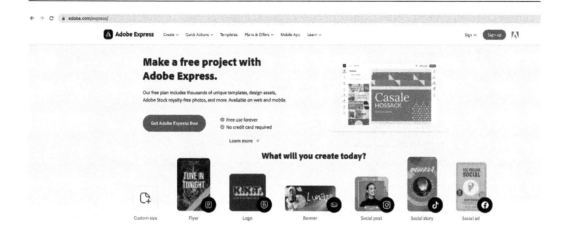

Figure 6.37 – Accessing Adobe Express via the browser

2. Navigate to the + icon.

Figure 6.38 – Click on the + icon to open the Quick Actions window

3. When you click on the + icon, the **Quick actions** tab opens.

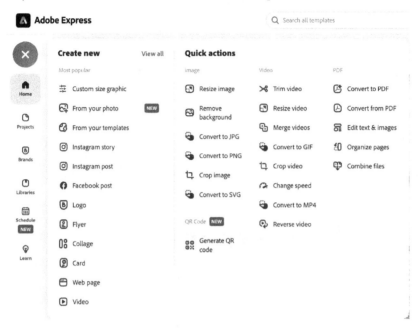

Figure 6.39 – Quick actions options

4. Under **Quick actions**, navigate to the **Convert to JPG** option.

Figure 6.40 – Convert to JPG in Quick actions

5. When you click on **Convert to JPG**, Adobe Express will open the **Convert PNG to JPG** window.

Figure 6.41 – Convert PNG to JPG

6. You can either drag and drop an image or click on **Browse on your device**. For this example, I clicked on the browse option to open the upload window. Locate your image and then click on **Open**.

Figure 6.42 – Uploading an image from your device

7. Adobe Express will display the window again with the processed image. The window will display the following message to indicate the image has been processed: **Your JPG is ready to download**. Click on the **Download** button to retrieve your JPG file.

Figure 6.43 – Image has been converted into a JPG

8. Once your image has been downloaded, Adobe Express will display a notification saying **Your image has been downloaded**.

Figure 6.44 – Downloaded image notification

You have now successfully downloaded a JPG copy of your image.

Figure 6.45 – JPG copy of the image

In this section, we acquired the skill of converting an image from a PNG format into a JPEG format. With a diverse range of file formats available, Express simplifies the process of converting common file formats, such as from PNG to JPG, and vice versa. This is made possible by leveraging the power of Adobe's core photo-editing software, Photoshop, which is seamlessly integrated into Express.

Summary

In this chapter, you acquired valuable knowledge on utilizing some of the Quick Actions in Adobe Express, which included removing a background, cropping and resizing an image, and finally, converting a PNG file into a JPG file. These Quick Actions are powered by Adobe's powerful core apps, such as Photoshop, Premiere Pro, and Acrobat. By harnessing the power of Adobe's robust core applications, accomplishing such actions becomes a seamless process, requiring only a few clicks to achieve the desired results.

In the next chapter, we will delve into the extensive capabilities of Acrobat. You will discover how to convert a PDF into a JPG, combine PDF files, and edit text and images. These powerful Acrobat features will equip you with the necessary tools to efficiently handle various PDF-related tasks and enhance your document management workflow.

7
Polishing PDFs Using Quick Actions

Powered by Acrobat, Adobe's world-standard PDF software, you can now access powerful features right inside Adobe Express, using **Quick Actions**.

Tasks such as converting a PDF into a JPG (and vice versa) and organizing and combining PDF files are just a few of the use cases for the PDF editing tools you can utilize in Adobe Express.

We will cover the following topics in this chapter:

- How to convert files to/from PDF
- How to edit text and images
- How to organize and combine PDF files

By the end of this chapter, you will be able to use the **Quick Actions** tools for PDF files in Adobe Express. You will be able to convert files to PDFs and convert PDF files to other formats, you will be able to edit text and images in a PDF, and you will also be able to organize and combine your PDF files with ease.

How to convert files to/from PDF

To get started, follow these steps on the browser:

1. Navigate to the home page on your desktop, `https://express.adobe.com/`:

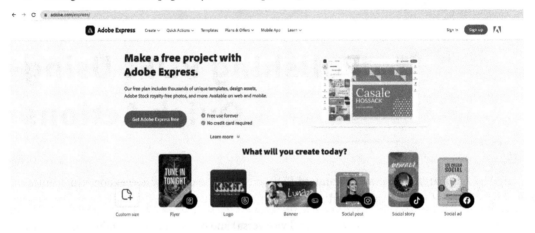

Figure 7.1 – Accessing Adobe Express via the browser

2. Navigate to the + icon:

Figure 7.2 – The + icon to open the Quick actions tab

3. Click on the + icon, which opens up the **Quick actions** tab:

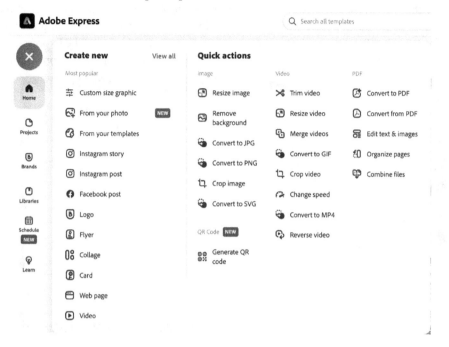

Figure 7.3 – Quick actions options

4. Navigate to the **PDF** section on the right and click on **Convert to PDF**:

Figure 7.4 – Convert to PDF quick action

5.   When you click on **Convert to PDF**, a window will appear. As you can see in the following screenshot, this **quick action** feature is powered by Adobe Acrobat. The file types you can convert into a PDF include Microsoft Word documents, Excel spreadsheets, PowerPoint presentations, and image files (JPG/PDF). There are two ways to import your file to Adobe Express. You can either drag and drop a file into the dotted rectangle or you can click on **Browse on your device**:

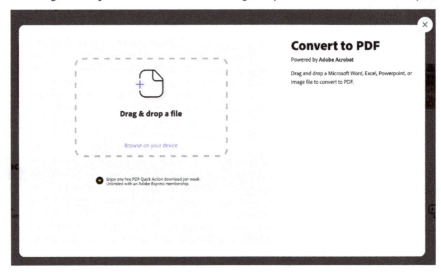

Figure 7.5 – Convert to PDF window

6.   When you click on **Browse on your device**, your finder image on Mac or folder on Windows will appear. Navigate to the file that you want to open. In this example, I am going to select a JPEG image file (.jpg). Next, I will click **Open**.

St Patrick's Day BBQ.jpg
JPEG image - 3.7 MB

Figure 7.6 – Navigate to your .jpg file on your device

7. Adobe Express will process this action:

Figure 7.7 – Adobe Express will process this image and convert it into a PDF

8. Once Adobe Express has finished processing the conversion of your JPEG file into a PDF, the **Download** button will appear:

Figure 7.8 – The image file has been successfully converted into a PDF

9. Click on **Download** to download your PDF. Adobe Express will display a window to notify you that your PDF has been successfully downloaded:

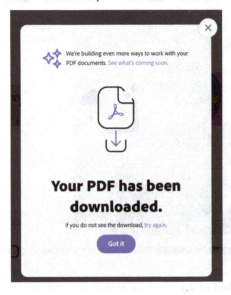

Figure 7.9 – Successfully downloaded notification

10. Simply click on **Got it** to exit out of this window.

11. Navigate to the `Downloads` folder on your hard drive to view your PDF:

Figure 7.10 – You now have a PDF version of the file

You have successfully converted your JPEG file into a PDF file:

Figure 7.11 – PDF file opened on Adobe Acrobat

In this section, we have acquired knowledge on how to convert a poster from JPEG format into PDF format. This powerful feature enables you to unlock the document's contents once it has been converted – a topic we will explore further in the next section.

In the next section, we will explore the process of editing text and images within a PDF document in Express.

How to edit text and images

To get started, follow these steps on the browser:

1. Navigate to the home page on your desktop, `https://express.adobe.com/`:

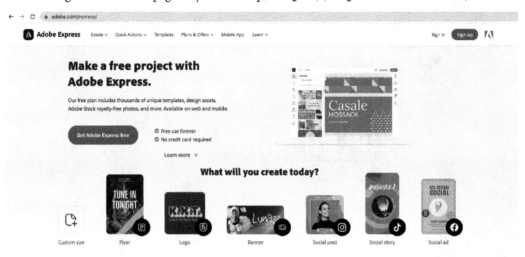

Figure 7.12 – Accessing Adobe Express via the browser

2. Navigate to the + icon:

Figure 7.13 – The + icon to open the Quick actions window

3. Click on the + icon, which opens up the **Quick actions** tab:

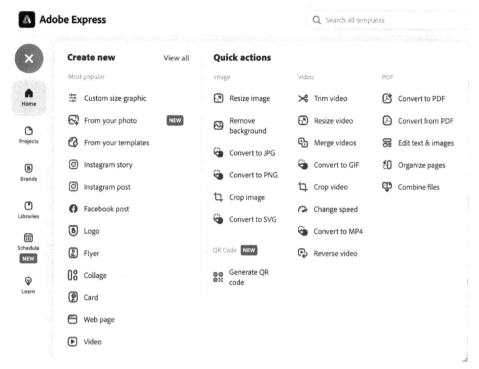

Figure 7.14 – Quick actions options

4. Navigate to the **PDF** section on the right and click on **Edit text & images**:

Figure 7.15 – Edit text & images quick action

5.   When you click on **Edit text & images**, Adobe Express will open a new window. You can either drag and drop your PDF or click on **Browse on your device** to upload it this way:

Figure 7.16 – Edit text & images window

6.   When you click on the option to browse, another window will open, and you can navigate to your file. Simply select your PDF, then click on the **Open** button:

Figure 7.17 – Browse through your folder to locate the PDF you want to upload

7. Adobe Express has successfully converted the PDF file and has recognized the text and images with its machine learning capabilities. Notice the marching ants around the images and text – this indicates that the content is now editable:

Figure 7.18 – Adobe has used machine learning to identify the text and images

8. For example, you can click and drag the image to reposition it. You can also rotate the image using the rotate tools in the toolbar.

Figure 7.19 – Select the image to reposition

When you select an item such as the image, the toolbar will appear. Formatting options for an image include rotating clockwise, rotating anti-clockwise, and deleting.

Figure 7.20 – Toolbar with image options

9. To edit any of the text, simply click on the text to replace the existing text:

Figure 7.21 – Highlight the text to edit

10. The toolbar will also show text formatting options, which include the flexibility to change the font, font size, font color, bold, italic, underline, and text alignment:

Figure 7.22 – Text formatting options available on the toolbar

11. Once you have made your edits, simply click on the **Download** button to download the updated version of your PDF file:

Figure 7.23 – Download button

12. Once your PDF has been downloaded, Adobe Express will display this notification window:

Figure 7.24 – Successful download notification

In this section, we have acquired the skills to perform PDF editing within Express. Typically, to edit a PDF, you would require access to a PDF editing software such as Acrobat. However, with Express, you can now access Acrobat's powerful editing capabilities directly within Express. Whether you need to modify event details on a flyer or edit a marketing proposal, you can now quickly perform these actions in Express.

In the next section, we will explore the additional PDF functionalities available in Express. We will explore the process of organizing and combining PDF files.

How to organize and combine PDF files

To get started, follow these steps on the browser:

1. Navigate to the home page on your desktop, `https://express.adobe.com/`:

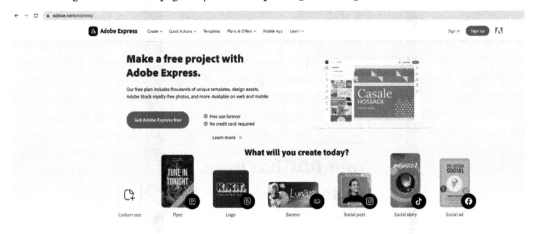

Figure 7.25 – Accessing Adobe Express via the browser

2. Navigate to the + icon:

Figure 7.26 – The + icon to open the Quick actions window

3. Click on the + icon, which opens up the **Quick actions** tab:

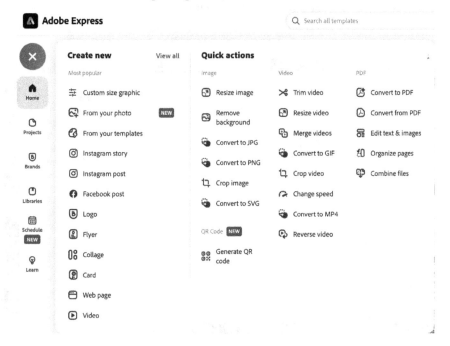

Figure 7.27 – Quick actions options

4. Navigate to the **PDF** section on the right and click on **Combine files**:

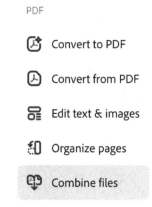

Figure 7.28 – Combine files quick action

5.   When you click on **Combine files**, a window will appear. You can combine the following file formats to combine into one PDF: PDF, Microsoft Excel, Microsoft PowerPoint, PNG, JPG, Rich Text Format, and Microsoft Word documents. Simply drag and drop files or click on **Browse on your device**:

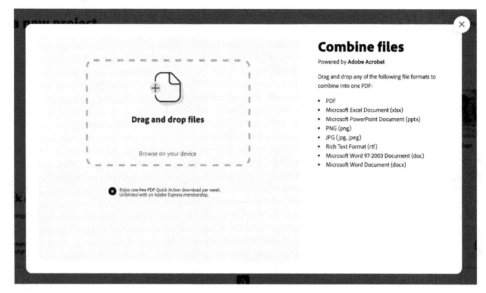

Figure 7.29 – Combine files window

6.   When you click on the option to browse, the folder window will open, and you can navigate and select multiple files. In this example, I am selecting two .pdf files and a .jpg file. I will then click **Open**:

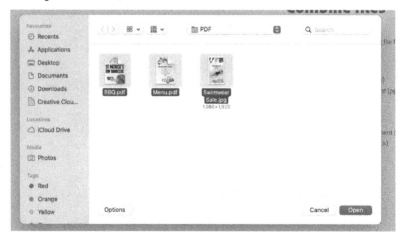

Figure 7.30 – Browse your folder and select the files you want to combine

7. Once your files have been uploaded to Express, the window will now display the files you have imported:

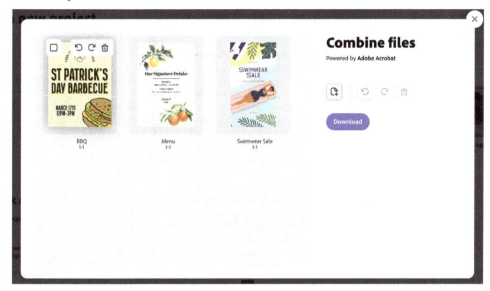

Figure 7.31 – Files you have uploaded will appear in the window

8. Click on **Download** to download the combined PDF file:

Figure 7.32 – When you click on Download, Express will start downloading the PDF

9. Once your file has been successfully downloaded, Express will display the following notification window:

Figure 7.33 – Successful download notification

10. Locate the file in your `Downloads` folder and open the combined PDF in Acrobat:

Figure 7.34 – Combined PDF

In this section, we have acquired the skills to combine multiple different file formats to create a single PDF file. Regardless of the file format (whether it is a Microsoft Excel file, Microsoft Word file, or JPEG file), you can combine all these different file formats into a single PDF document.

Summary

In this chapter, you have acquired proficiency in utilizing the PDF quick actions, which include tasks such as converting files to/from PDFs, editing text and images, and organizing and combining PDF files. Powered by Adobe's powerful PDF software, you can easily access these powerful features directly within Express.

In the next chapter, we will embark on our mini-projects. The projects we will be creating in the next chapter will include creating an animation for an Instagram story, creating a marketing campaign, and finally, creating an event poster.

8
Put Your Skills to Practice with Adobe Express

In this chapter, you will be guided through three practical exercises to put your skills to practice. These exercises will help you to learn your way around Express. You will learn design techniques and discover ways to use the various functions in the application. This will provide you with the confidence to explore and start creating your own projects, where you can incorporate the design techniques you have learned.

We will cover the following topics in this chapter:

- Creating an Instagram story from scratch
- Creating a marketing campaign
- Creating an event poster

By the end of this chapter, you will know how to use layers and the design assets and images in Express to create dynamic designs that would impress even professional graphic designers. Express has integrations with Adobe's core apps, including Acrobat, Photoshop, and Premiere Pro, which unlocks endless possibilities for what you can create in Express. Functions such as the **Remove background** tool allow you to create dynamic composites quickly and easily. You can intertwine images with text to create 3D designs, elevating otherwise-boring two-dimensional creations. Keep on reading, and better yet, grab your laptop or mobile device to follow along and start creating.

Creating an Instagram story from scratch

For the first project, we will be creating an Instagram story for a fictitious fashion e-commerce company called *Marley OOTD* (**OOTD** stands for **outfit of the day**). We will add a design asset, **Dots**, as the backdrop. We will find an image from Adobe Stock and remove the background. We will then duplicate the masked image and create multiple layers. We will add text and then incorporate the masked image with the text, where we will have the model's legs looping through some of the letters, to create a 3D illusion. The following figure shows what the final image will look like:

Figure 8.1 – Final image

To get started, follow these steps in the browser:

1. Navigate to the home page of Adobe Express on your browser at `https://express.adobe.com/`.

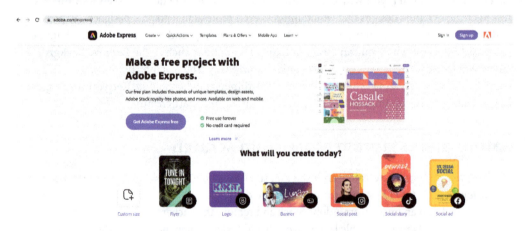

Figure 8.2 – Access Adobe Express via the browser

2. Once you are logged in, click on the + button on the top left to start a new project.

Figure 8.3 – Start creating by clicking on the + button

3. Next, click on **Instagram story**.

Figure 8.4 – Choose one of the options to start creating

Express will automatically open a new blank project in the dimensions of an Instagram story.

4. To start, click on **Photos** and type in a keyword to search images from Adobe Stock. I typed the words `fashion heels` and then clicked on the image. Express will then add this image to the project.

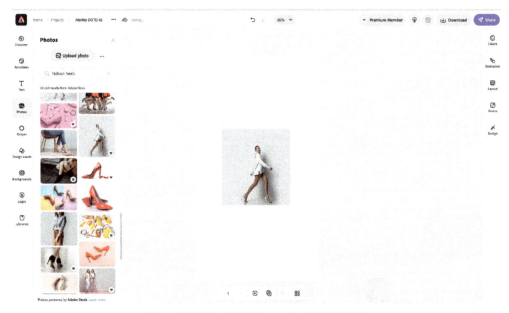

Figure 8.5 – Search for images in Adobe Stock

5. Next, click on the image. The image options window pane will pop out from the right-hand side of the page. Click **Add to background** to add the image to the background:

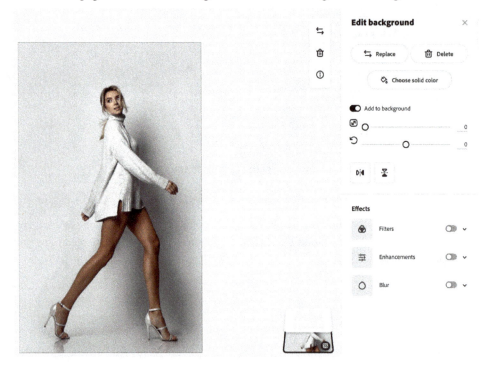

Figure 8.6 – Click on the image, then click on the Add to background button

6. Next, click on **Design assets** on the left. Navigate to **Textures** by clicking on the more button, then navigate through the textures and click on **Dots**:

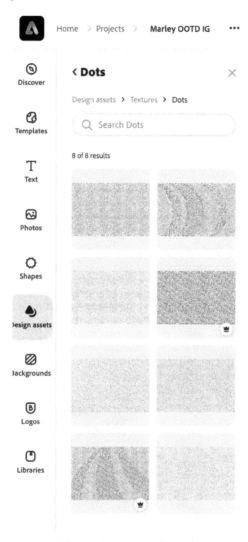

Figure 8.7 – Design assets > Textures > Dots

7.   Click on one of the dot textures. Express will automatically add this to your project.

Figure 8.8 – Express will add the texture to your project

8.   Click on the image and resize it to the size of your vertical artboard (click on the circles and drag them out).

Figure 8.9 – Resize the dots texture so it covers the entire artboard

9. Next, click on the dots to open the layers (located at the bottom right).

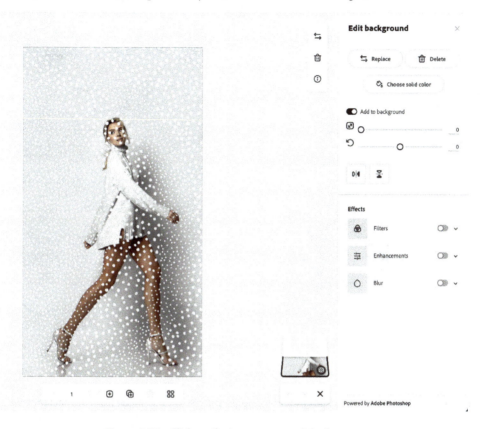

Figure 8.10 – Click on the image to reveal the layers

10. Click on the image of the model layer, then deselect **Add to background**. This will release the layer of the model, so we can duplicate this layer:

Figure 8.11 – Click on Add to background to toggle and release the image from the background

11. Click on the duplicate layer icon to duplicate this layer containing the model:

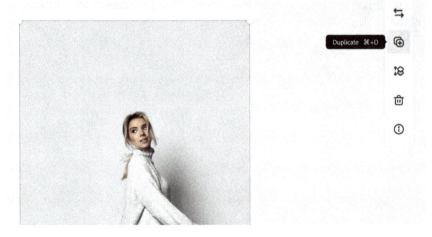

Figure 8.12 – Duplicate the layer

There are now two layers of the model:

Figure 8.13 – Two layers displayed in the layer stack as well as the project

12. Click on the bottom layer of the model layer, then click on **Add to background** to add the model back to the background layer:

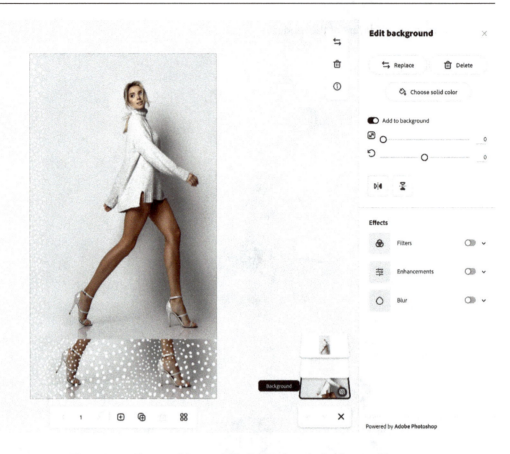

Figure 8.14 – Fix one of the model layers back to the background layer

13. Next, click on the top model layer, then click on the **Remove background** button:

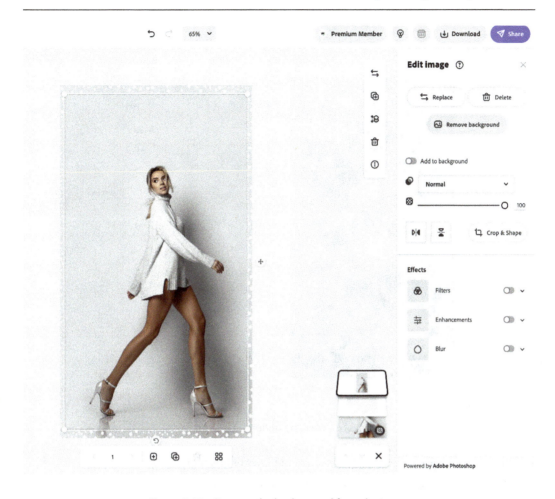

Figure 8.15 – Remove the background from the image

14. Adobe Express will remove the background. Click on the tick icon to accept the cutout:

Figure 8.16 – Masking options in Express

15. Resize the image so it fits over the background layer of this same image.

16. Next, click on **Text**, then click on the **Add your text** button.

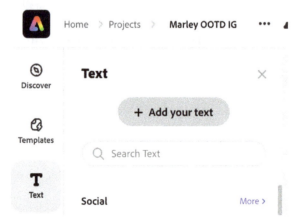

Figure 8.17 – Add text to your project

17. Override the text by typing in OOTD, and change the font to **Juniper Std Medium** and color to **#E44418**, or select your preferred color for this project. Resize the text and move it to the bottom of the screen:

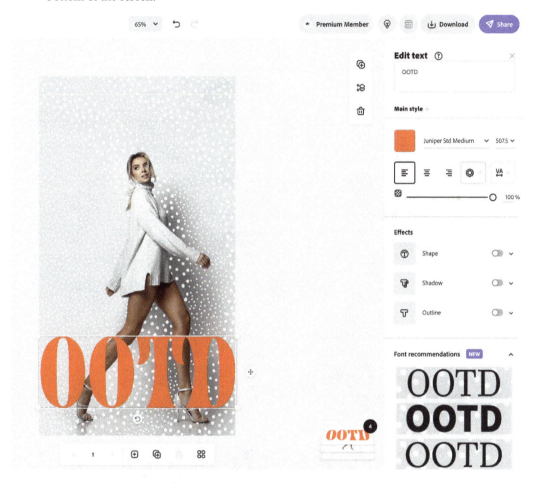

Figure 8.18 – Format the text

18. Next, change the text alignment to **Semi-circle** by clicking on the donut icon

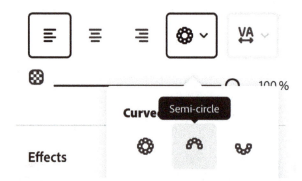

Figure 8.19 – Change the text alignment

19. Resize and position the text over the model's legs, like the following.

Figure 8.20 – Position the text over the image

20. Now, we will start creating duplicates of the model so we can intertwine her legs with the text. To do this, click on the model, then click on the duplicate icon to create a duplicate of this layer (refer to *step 11*). Then, click on the new duplicated layer and move it to the top so it is above the text layer:

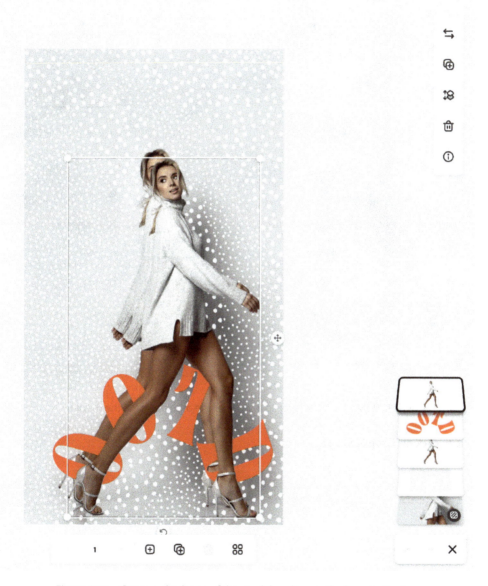

Figure 8.21 – Create a duplicate of the model and move the layer to the top

21. Position the model so that the layer sits on top of the other layers neatly:

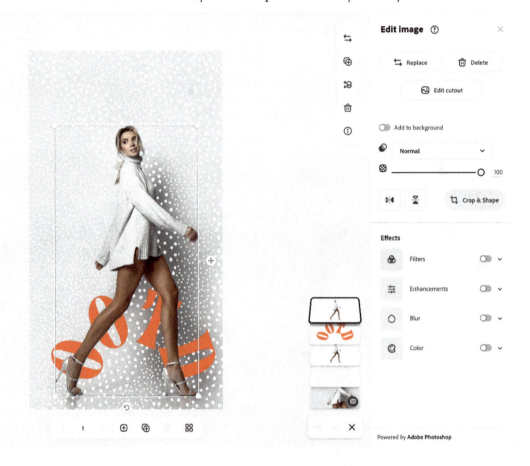

Figure 8.22 – Position the layer so it sits over the other layers of the model

22. Next, click on **Crop & Shape**:

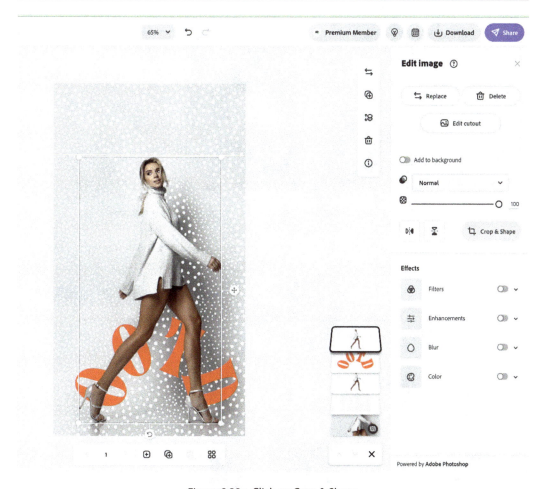

Figure 8.23 – Click on Crop & Shape

23. Click on **Freeform**, then drag the corners so you have part of the leg over the second letter (second **O**). This gives the illusion of the leg going through the letter **O**. Click on the tick to accept this change. You may need to move the text layer to get this precise mask:

Figure 8.24 – Crop the mask so we have this part of the leg in the top layer

24. Next, repeat the process of duplicating the model layer and cropping the new layer to position the leg going through the first letter. See the following figure for reference:

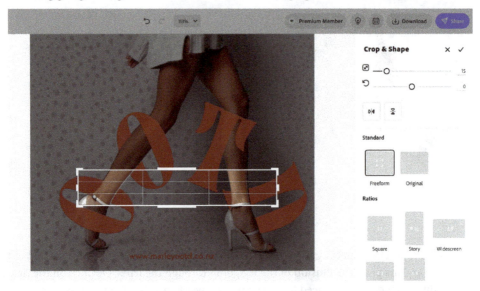

Figure 8.25 – Crop the second mask to loop over the first letter

25. Repeat the process and duplicate yet another layer. Crop the right leg so it's in front of the last letter, **D**.

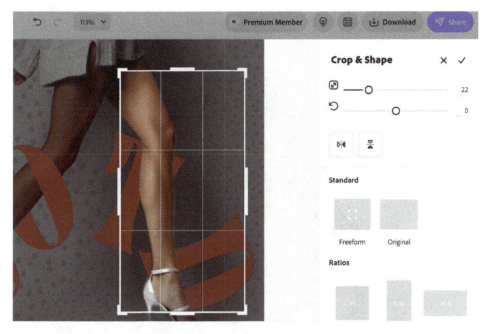

Figure 8.26 – Crop the new layer so the right leg is over the letter D

26. Next, duplicate the word, but we only need the letter T. Position this letter so it sits over the model's leg:

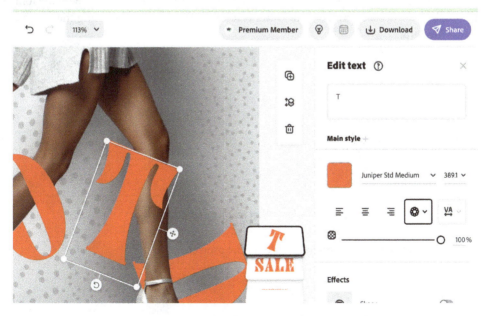

Figure 8.27 – Duplicate the word and position the letter T over the model's leg

Your project should look as in the following figure so far:

Figure 8.28 – What your project should look like so far

27. Next, add any other text to your project. I added the word **SALE** and used the same font as the **OOTD** word. I also added some other text below **SALE**, this time using the font **Neue Kabel Medium**.

Figure 8.29 – Add other text to your project

28. Add the website URL to the bottom center of the project.

Figure 8.30 – Add the website/call to action

Congratulations, you have successfully created your first project in this chapter! Your Instagram story project should look similar to the following:

Figure 8.31 – End result

In this section, you learned how to create a photo composite with text. You learned how to remove a background from an image and create multiple copies of that masked image and utilized layers to intertwine the image with the text.

With this technique, you now have the technical skills to create professional 3D photo composites and start to create creative complex compositions.

In the next section, we will learn how to create a marketing campaign that can be repurposed for a website splash page or used on social media.

Creating a marketing campaign

In this section, we will learn how to create an eye-catching marketing campaign for a fictitious fitness brand. We will learn how to use design assets in Express, create a photo composite, and use text templates.

The following figure shows what the final result will look like:

Figure 8.32 – Final image

To get started, follow these steps in the browser:

1. Navigate to the Adobe Express home page on your browser at `https://express.adobe.com/`.

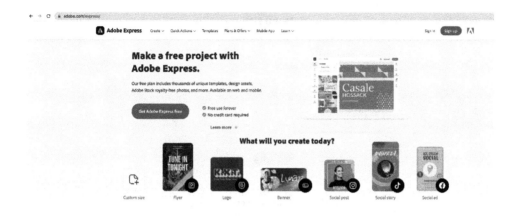

Figure 8.33 – Access Adobe Express via the browser

2. Select the + icon to start a new project:

Figure 8.34 – Start creating by clicking on the + button

3. Next, click on **Custom graphic size**, select **YouTube thumbnail** under **SOCIAL POST**, then click on **Next**. Express will take you to a new blank project:

Figure 8.35 – Choose a custom size

4. Next, click on **Design assets** and type in the `stripe` keyword to find a design asset you like. Click on the thumbnail to add this to your project and resize it to fit:

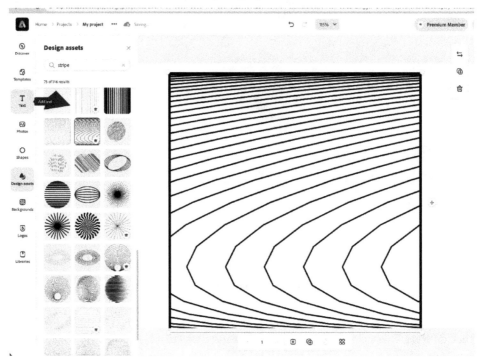

Figure 8.36 – Type in a keyword to find design assets

5. Next, click on **Photos** and type in the iphone mockup keyword. Find a suitable image to use in this project. Click on the thumbnail image of your selection of a phone mockup:

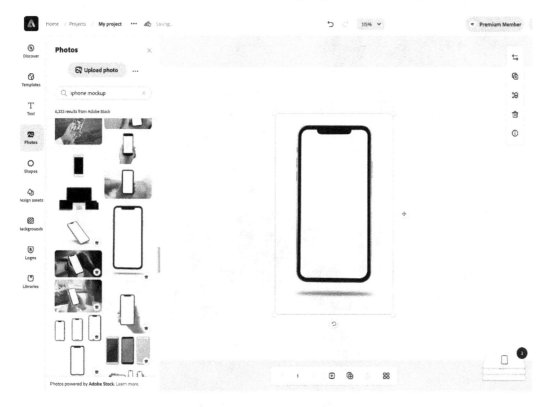

Figure 8.37 – Search for an image in Adobe Stock

6. Click on the image, then click on **Remove background** so we have a mask of the iPhone only. Click on the tick icon to accept the mask:

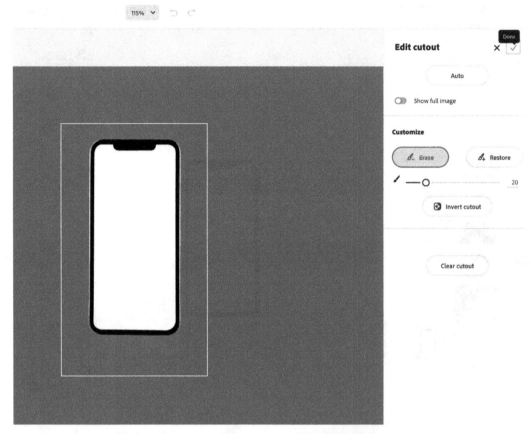

Figure 8.38 – Remove the background from the phone mockup

7. Next, repeat the process and find an image of a fitness model. The keywords I used were `fitness girl`. When you click on your selection, Express will add this to your artboard.

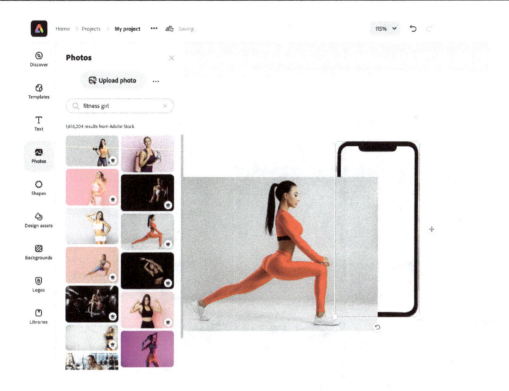

Figure 8.39 – Find an image from Adobe Stock

8. Remove the background from the image and position the image inside the phone.

Figure 8.40 – Position the fitness model inside the phone mockup

9. Next, create a duplicate image of the model and position this duplicate layer over the original layer.

10. Click on the top layer and click on **Crop & Shape**, then click on **Freeform** and crop the image so the model is flush with the phone mockup:

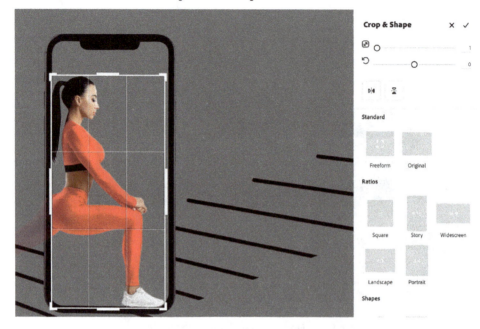

Figure 8.41 – Crop and position the layer

11. For the other layer of the model, crop the image so her legs are flush with the outside of the phone mockup:

Figure 8.42 – Crop and position the other layer

This will give the illusion of the model's leg stretching out of the mockup.

Figure 8.43 – Model appears to be stretching out of the frame

12. Next, click on **Text**, navigate to **Phrases**, and select one of the text templates. You can format the text on the right-hand side. I changed the text to TRAIN WITH ME and the color of the outline to red. I also rotated the text. The font I used was **League Gothic Italic**.

Figure 8.44 – Add text from a text template and customize it to your liking

13. Next, repeat the previous step, choosing a text template from the **Lists** category. Replace the text and change the formatting to your liking.

Figure 8.45 – Add more text to your project

14. Next, add the remainder of the text to this project:

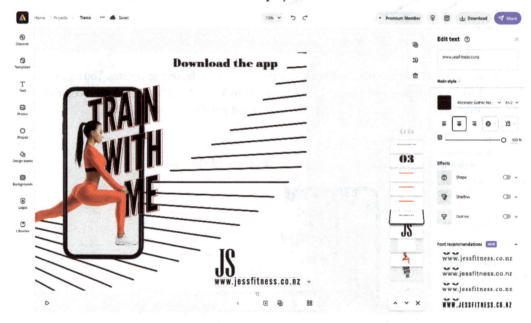

Figure 8.46 – Add the rest of the text to your project

Congratulations, you have successfully finished your second project! You should be proud of yourself. The following figure displays what your project may look like:

Figure 8.47 – End result

Bonus tip

Click on **Animation** to animate the text or images in your project:

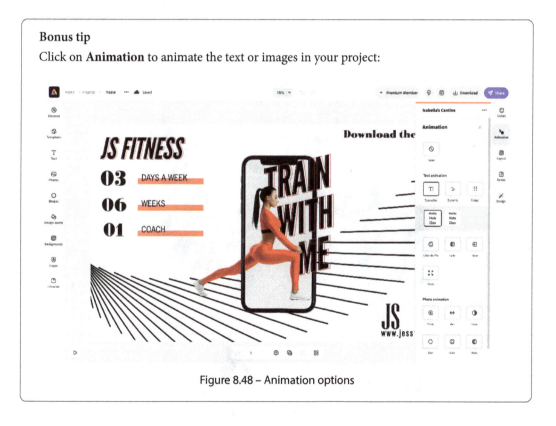

Figure 8.48 – Animation options

In this section, you learned how to create an eye-catching marketing campaign. You can use this content for a website splash page, repurpose it for a YouTube thumbnail, or animate the content for social media. The techniques you learned in this section will help you think outside the box (literally) and see how you can use different elements to create a dynamic composition for any campaign.

In the next section, you will learn how to create an event poster.

Creating an event poster

In this section, you will learn how to create an event poster that can be printed or used digitally. The skills you will learn in this section include how to create a photo composite using different elements, such as an image combined with an illustration. You will also learn how to utilize the text alignment formatting option, called magic, which will elevate your text composition and overall design aesthetic.

This will be the end result of your final project in this chapter:

Figure 8.49 – Final result

To get started, follow these steps in the browser:

1. Navigate to the Adobe Express home page on your browser at `https://express.adobe.com/`.

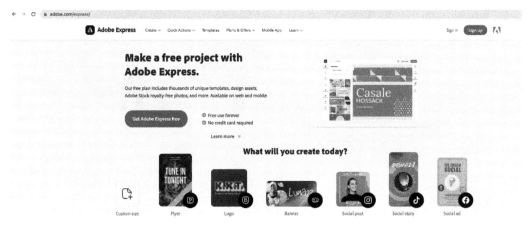

Figure 8.50 – Access Adobe Express via the browser

2. Select the + icon to start a new project:

Figure 8.51 – Start creating by clicking on the + button

3. Next, click on **Custom graphic size**, then select **Poster** from the **PRINT** category. Then, click on **Next**. Express will take you to a new blank project.

Figure 8.52 – Choose a custom size

4. Click on the canvas and change the background color using this hex code: `#ECF0F199`. Otherwise, feel free to pick your own color:

Figure 8.53 – Change the background color

5. Next, click on **Photos** and type in the keywords `girl face`. Select an image you want to use in this project. Click on the thumbnail of the image and Express will add it to your project:

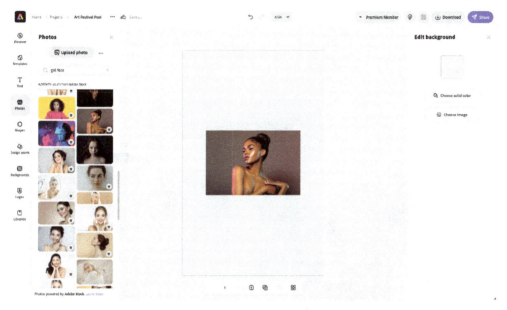

Figure 8.54 – Browse through millions of images on Adobe Stock

6. Click on the image, click on **Remove background**, then click on the tick icon to apply the mask:

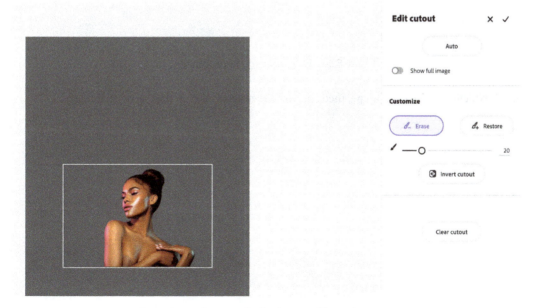

Figure 8.55 – Remove the background from the image

7.   Next, resize and flip the image horizontally:

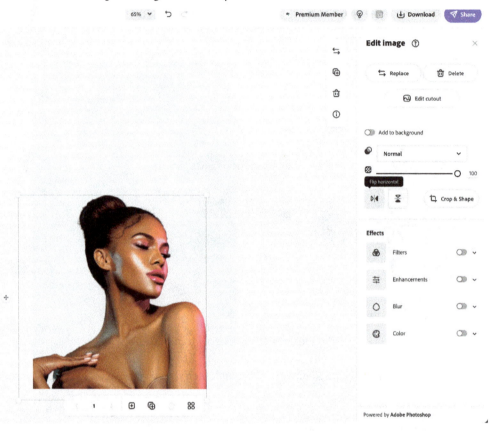

Figure 8.56 – Flip the image horizontally

8.   Next, click on **Crop & Shape**, then click on **Freeform** and crop the top of the image:

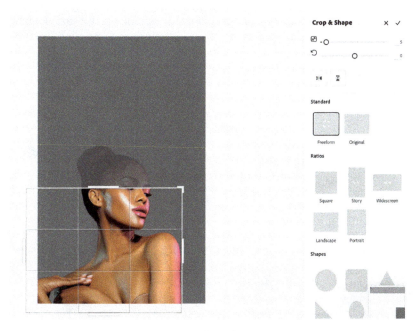

Figure 8.57 – Crop the image

9.   Next, click on **Photos** and search for `flower illustration` in Adobe Stock. Click on an illustration to add it to your project:

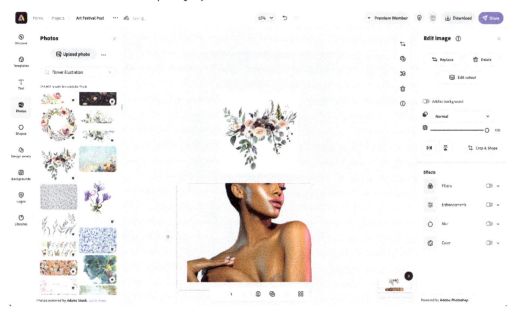

Figure 8.58 – Search for a flower illustration from Adobe Stock

10. Click on the image, click on **Remove background**, then click on the tick icon to apply the mask:

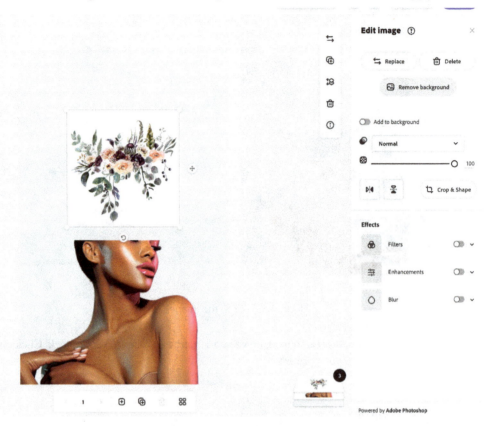

Figure 8.59 – Remove the background from the flower illustration

11. Click on **Flip horizontal** and place the illustration over the model's face:

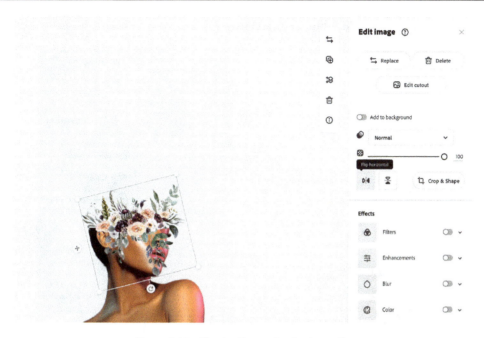

Figure 8.60 – Flip the illustration horizontally

12. Click on the illustration and create a duplicate. Rotate the second illustration and stack it over the first illustration:

Figure 8.61 – Stack the second illustration on top of the first illustration

13. Repeat this process to add a third illustration:

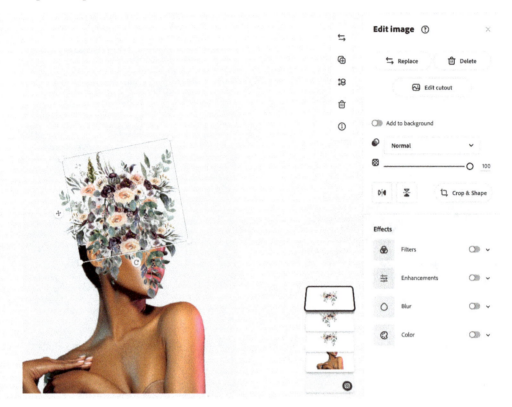

Figure 8.62 – Duplicate the illustration again

14. Next, click on **Text**, then **Add your text**. Type LIFE ISN'T SERIOUS in the **Edit text** field and change the font to **Peacock Deep Bold**. Change the font color to the #73947A hex color:

Figure 8.63 – Add text to your project

15. Next, click on **Shape**, then select the circle shape. Change the shape color to the #19291D hex color and drop the opacity down to **70%**.

Figure 8.64 – Add a shape to your text

16. Next, change the text alignment to **Triplet**.

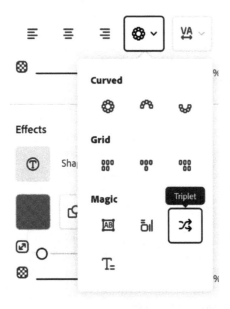

Figure 8.65 – Change the text alignment

17. Next, add a shape to the text. I picked a circle shape:

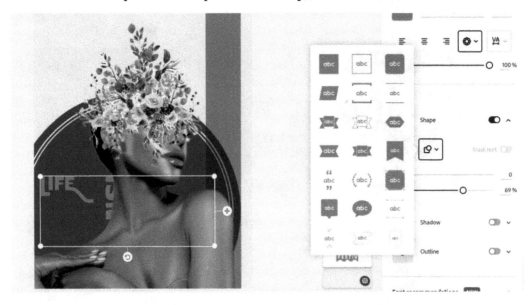

Figure 8.66 – Add a shape to the text

18. Next, click and drag the text layer so it is underneath the model layer. We want the image to be in front of the text layer.

Figure 8.67 – Reposition this layer

19. Next, duplicate the text layer, click on **Outline**, toggle **Transparent text** on, and change the thickness to 1. Then, drag this new text layer so it is on top of the image layer. This gives the illusion of the text blending over the model, yet, transitions to an outline only when the text is on the model's body:

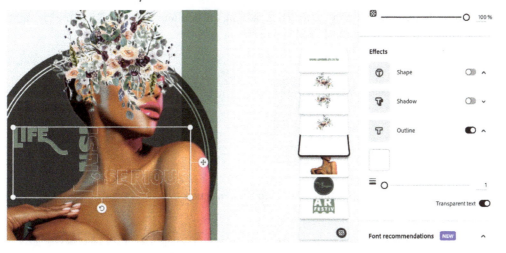

Figure 8.68 – Duplicate the text and sandwich the image between the two text layers

20. Next, add a new text layer, change the text to Art Festival 2025, and change the alignment to **Capitalize & fit**:

Figure 8.69 – Add a new text layer

21. I chose the font **Source Sans Pro Black** and changed the font size to **53**. Next, add a shape and change the color to this: #73947A. Next, resize and reposition the text so it sits in the top-right corner.

22. To finish the design, I added a call to action, adding the fictitious website URL to the top right. The font I used is **Amboy Black**, with a font size of **30.9**:

Figure 8.70 – Add a call to action

Congratulations, you have successfully completed your final project in this chapter! This is what your final project may look like:

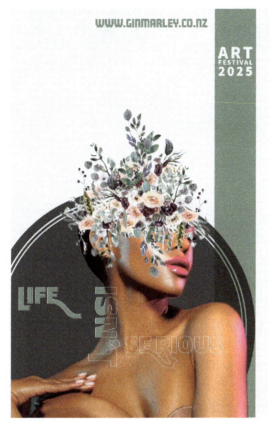

Figure 8.71 – Final result

In this section, you learned how to create an event poster. You have picked up even more skills in this exercise, such as sandwiching an image between two text layers.

Summary

In this chapter, you accomplished a major milestone by completing three exercises that have taught you crucial graphic design skills. Unlike previous chapters, where you utilized the pre-made templates in Express, this chapter encouraged you to create projects from scratch and take full control over the creative process, allowing you to master new techniques along the way.

By completing the exercises in this chapter, you are now equipped with a comprehensive understanding of a diverse range of graphic design techniques. In this chapter, you have gained the skills to create photo composites, by intertwining images with text to create a 3D image. Other techniques you have acquired include the ability to blend mixed media to create eye-catching visuals and utilize layers to produce dynamic and visually appealing designs. These are all highly desirable skills in graphic design, which you have accomplished in this chapter; you can be proud of the newfound expertise you have acquired.

In the next chapter, you will learn how to create a one-page no-code web page. No coding skills are required to create a dynamic and responsive web page in Express.

Part 3 – Create a Web Page with Adobe Express

In this part, you will learn how to create a professional one-page website without coding, featuring images, videos, formatted text, and shareable for free. The guide also includes practical exercises and examples to help you create engaging and effective web pages for various purposes, all hosted on Adobe's platform.

This part has the following chapters:

9

Building a Web Page with Adobe Express

In this chapter, you will acquire the skills to create a single-page website without any prior coding knowledge. Adobe's platform allows you to host these web pages at no cost, and sharing them publicly is a seamless process. Web pages are ideal for various purposes, such as marketing splash pages, presentations, blog posts, portfolios, resumes, registration pages, newsletters, advertisements, and product catalogs.

In Express, creating a single-page website has never been simpler. Not only is it a straightforward process but you also achieve a professional-looking web page as the result.

This chapter will cover the following topics:

- How to create a web page – adding images and videos
- Adding and formatting text and changing the theme
- Sharing your web page

By the end of this chapter, you will have the ability to swiftly and easily create a responsive web page. You will be able to create a web page within minutes.

How to create a web page – adding images and videos

To get started, follow these steps within your web browser:

1. Access the home page on your desktop by visiting `https://express.adobe.com/`.

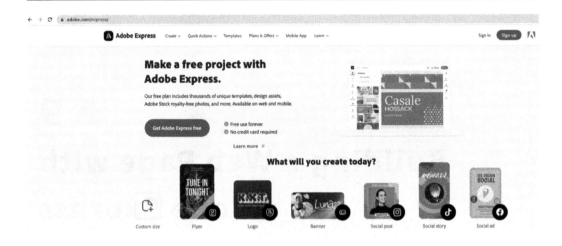

Figure 9.1 – Access Adobe Express via the browser

2.   After logging in, locate and click on the + button to create a new project.

Figure 9.2 – Start creating by clicking on the + button

3.   In the **Create new** section, click on **Web page**.

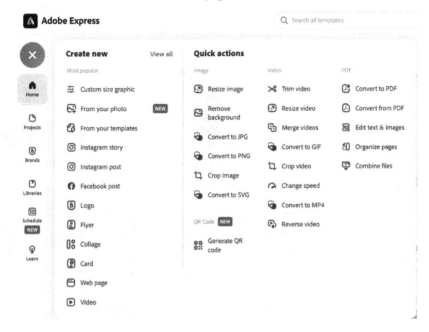

Figure 9.3 – Create new web page

4.   Adobe Express will create a new web page.

Figure 9.4 – New web page

5. Select the **Photo** option from the + button, then proceed to either upload your own image or explore the Adobe Stock library. Click on an image you wish to use, and Express will automatically add the image on the first page of your web page.

Figure 9.5 – Find an image from Adobe Stock

6. Next, replace the placeholder text currently displayed as **Add a title** and **Add a subtitle**.

Figure 9.6 – The placeholder text

7. Replace the text placeholder with your own text. For this exercise, I replaced the text with **JS Fitness** for the title and **Your personal fitness coach** for the subtitle.

Figure 9.7 – Replace the placeholder text

8.   To start building your web page, all you need to do is to scroll down to view the callout tab prompts located at the bottom of the page. You will be presented with layout choices, as per *Figure 9.8*. For this exercise, let's opt for the **Glideshow** layout, but feel free to explore the other layout options as well to enhance your practice.

Figure 9.8 – Prompt guides to add content

9.   Once you click on the **Glideshow** option, Express will open the **Photos** window, positioned on the right-hand side of the page, and showcase your most recent Adobe Stock search. Simply click on any desired photos to incorporate them into your web page. Finally, click on the **Save** button, located in the top-right corner, to add these images to your web page.

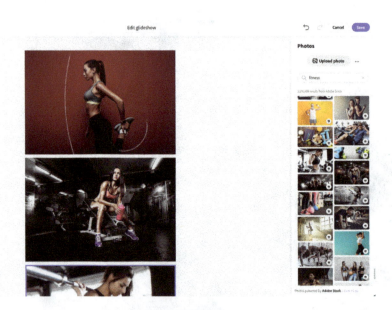

Figure 9.9 – Add images to your Glideshow

10. Once you have clicked on the **Save** button, Express will redirect you back to the web page. At this point, locate the + button within the semi-transparent rectangle and click on it. Express will then provide guidance for adding any of the following four elements: **Photo**, **Text**, **Button**, or **Video**.

Figure 9.10 – Add more elements to your web page

11. When you click on the **Photo** option, this will prompt Express to open the **Photos** window once again to the right of the page. Select your desired image and Express will automatically add the image inside the rectangle.

Figure 9.11 – Add an image from Adobe Stock

12. Within the current interface, you will notice two + buttons, one positioned above the image and the other beneath it. Proceed by clicking on the lower + button and selecting the **Video** option. You have the flexibility to add a link from YouTube, Vimeo, or Adobe Express.

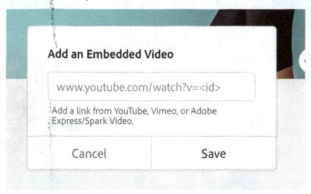

Figure 9.12 – Add an Embedded Video

13. Retrieve the URL link of the desired video you wish to embed, then paste the link to incorporate the video.

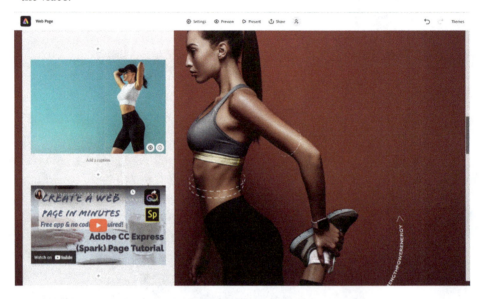

Figure 9.13 – Embed a video into your web page

14. To reposition the content box, you can effortlessly adjust its alignment by simply clicking and dragging it. You can align the content box to the left, center, or right of your web page based on your preference.

Figure 9.14 – Change the alignment of the rectangle content box

15. To proceed with adding more content, simply scroll down and select any of the available content options to continue building your web page. In this case, let's select the **Split layout** option.

Figure 9.15 – Add more content

16. When you select the **Split layout** option, Express will generate a fresh page featuring a vertical division down the center. On one side, you can easily include an image, while on the other side, you have the flexibility to add various content such as text. If you wish to swap the layout of the image and text, simply click on the arrows located in the middle of the page.

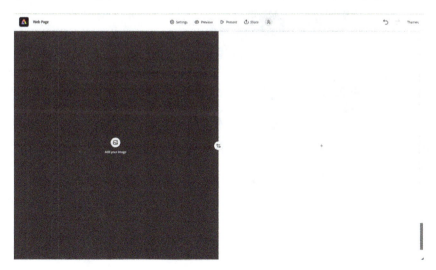

Figure 9.16 – Split layout

17. Click on **Add your image** and add an image from Adobe Stock.

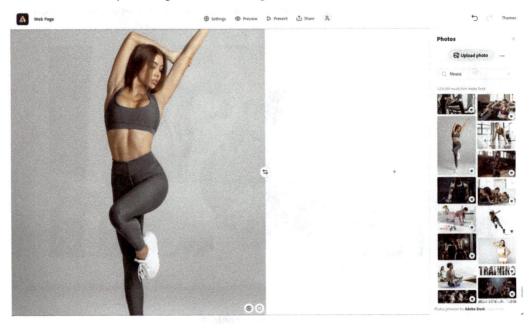

Figure 9.17 – Add an image from Adobe Stock

18. To adjust the focal point of the image, you can easily achieve this by simply clicking on the image itself. Express will then present a fly-out prompt containing the focal point display, where you can find a button labeled **Focal point**.

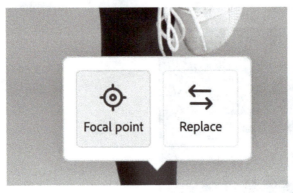

Figure 9.18 – Change the focal point

19. When you click on the **Focal point** button, you have the flexibility to adjust the focal point of the image by moving the circle accordingly. Additionally, the thumbnail located in the top-right corner provides a portrait preview of the changes. Once you are happy with the position of the image, click on the **Save** button to save your changes.

Figure 9.19 – Change the focal point by moving the circle

In this section, you have acquired the skills to create a web page in Express. You have also learned about the process of adding content and building your web page, including the incorporation of photos from Adobe Stock and displaying them in different styles such as full-screen and split layout.

In the next section, we will explore the process of adding and formatting text, as well as discovering the process of changing the theme of your web page.

Adding and formatting text and changing the theme

In this section, we will focus on adding text to a web page.

The content on your website should not only be engaging and informative for the audience but the formatting and layout should be given careful consideration too. To get started, follow these instructions:

1. Continuing on from the previous section, proceed by clicking on the + button located on the split layout page. Express will then present you with several options for adding content, including an image/photo, text, button, or video.

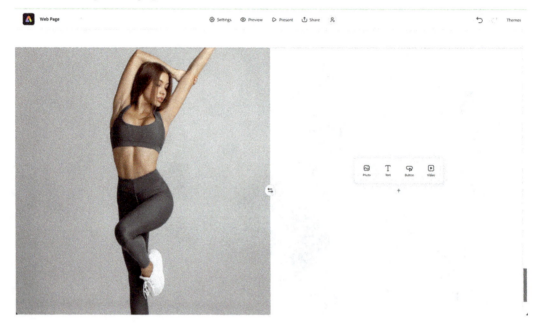

Figure 9.20 – Add more content to your split layout

2. For this exercise, select **Text** and proceed by adding your desired text.

3 Daily Goals

Figure 9.21 – Add your text

3. After you have added your desired text, simply click on the text you have inserted. Express will then present you with a range of text formatting options, allowing you to apply styles such as bold, italics, bullets, alignment, and even the inclusion of hyperlinks. This gives you the flexibility to customize and enhance the appearance of your text according to your preference.

3 Daily Goals

Figure 9.22 – Text formatting options

4.   For this exercise, let's opt for the **H1** or **H2** style. When you click on the text style, Express will automatically apply the chosen text style.

Figure 9.23 – Apply the H1 text style

5.   Below the title, click on the + button to insert additional text. To apply a bullet point style to your text, select the **bullet points** option.

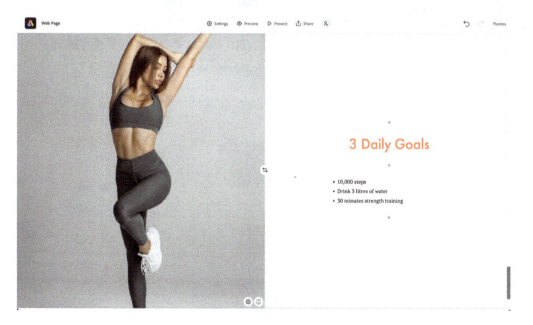

Figure 9.24 – Bullet style applied

6.   Currently, it is not possible to change the font in Express (as of the time of writing this book). However, you can change the overall theme, which will alter the font and listed attributes.

7. Proceed to click on the **Themes** option located in the top-right corner of the page. Express will then present you with a collection of pre-designed themes that alter the overall appearance of your web page, encompassing font styles and transitions. If you have already created a brand within Express, you will notice the presence of **Brand Themes**. Otherwise, if you haven't established a brand, Express will only display **Adobe Express Themes**.

Figure 9.25 – Change the theme

8. To see a preview of each theme's visual presentation, simply select and review them one by one.

Figure 9.26 – WESLEY theme applied

9. Scroll through your web page to see how the selected theme alters the font styles throughout the content. Upon clicking on a theme, Express will automatically apply the chosen theme to your web page. Once you have made your selection, click on **Themes** again to close the pane and finalize your theme choice.

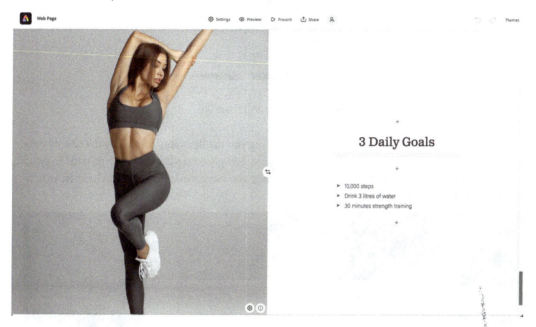

Figure 9.27 – Vintage theme applied

In this section, you have acquired the skills to add and format text on your web page. Additionally, you have learned how to change the theme, which affects the font, font size, and font color throughout the page.

In the next section, you will discover the process of sharing your web page.

How to share your web page

In this section, we will focus on learning how to share your web page. By default, Express web pages are hosted on Adobe Express.

To begin the sharing process, follow these steps within the browser:

1. Click on the **Share** option located at the top of the page. Express will present you with various sharing options, including **Publish and share link**.

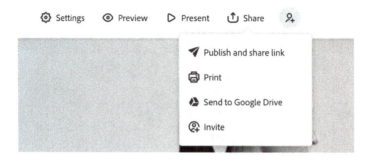

Figure 9.28 – Publish and share link

2. Express will display the **Publish** settings, granting you the flexibility to modify the title of your web page, change the category, display your name, add photo credits, and even provide Express with permission to feature your web page. Once you have made the desired changes, simply click on **Create link** to proceed.

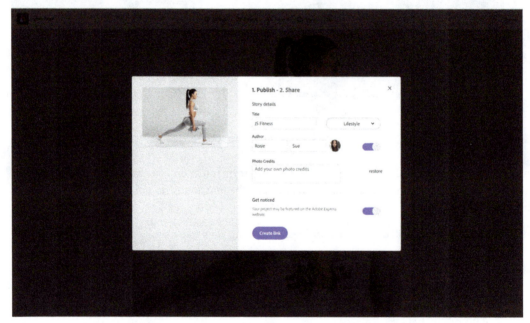

Figure 9.29 – Publish options

3. After generating the link, Express will provide you with a shareable URL that you can use to share your web page.

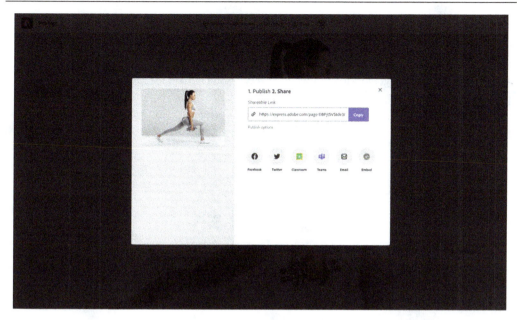

Figure 9.30 – Shareable link

4. Click on the **Copy** button to copy the URL of your newly created web page.

Figure 9.31 – Access your public-facing Express page via the URL

In this section, you have acquired knowledge on the process of creating a URL to publish and share your web page. By default, Express will host web pages on their server. However, if you wish to remove your web page from being publicly accessible, you can choose to unpublish it at any time. Simply navigate to the **Projects** page, locate the specific project, and click on the ellipsis icon to reveal the **Unpublish** option.

Figure 9.32 – Scan the QR code to view the web page created in this chapter

Summary

In this chapter, you have been introduced to the world of creating web pages using Express. Even if you have no prior experience or coding skills, Express provides an easy and accessible way to create stunning and professional web pages in a matter of minutes. With the help of intuitive prompts and guides integrated into Express, you can simply scroll and add content as you go, minimizing the chance of errors.

Throughout this chapter, you have acquired a range of skills related to web page creation. These include adding images and videos, formatting and modifying text, and customizing the theme of your page. Additionally, you have also been shown the process of sharing your web page.

This chapter has provided you with a valuable set of skills and knowledge that will enable you to create visually appealing and engaging web pages. Whether you are a beginner or an experienced web designer, the tools and techniques covered in this chapter will help you to create web pages that truly stand out.

In the next chapter, you can look forward to a hands-on approach to creating your own web page. The chapter will provide examples and practical exercises, offering you the opportunity to create different types of web pages.

10
Mini Projects – Creating Your Own Web Page(s) with Adobe Express

In this chapter, we will explore a step-by-step guide to creating web pages without any coding skills required. These web pages are hosted on Adobe's platform, at no additional cost, and can easily be shared with a public audience. This chapter highlights the versatility of web pages, as they can be utilized for various purposes, such as marketing splash pages, presentations, portfolios, resumes, registration pages, newsletters, advertisements, business proposals, and product catalogs.

Practical exercises and examples are included in the chapter to provide you with hands-on experience of creating different types of web pages. The two practical exercises included are as follows:

- Creating a portfolio as a **User-Generated Content Creator** (**UGC Creator**)
- Designing a marketing splash page for an e-commerce business

Through these exercises and examples, you will gain insights into the process of creating captivating and engaging web pages without having any coding knowledge. The web pages created are hosted on Adobe, enabling you to share them easily with others.

Creating a portfolio as a UGC Creator

A UGC creator portfolio is a web page that showcases content created by a user, such as photography or videos, along with a rate card. In this exercise, we will create a mock portfolio for a fictitious UGC creator, using the persona of Sofia Lopez.

To get started with the first project, follow these steps on your browser:

1. Navigate to the home page on your desktop: `https://express.adobe.com`.

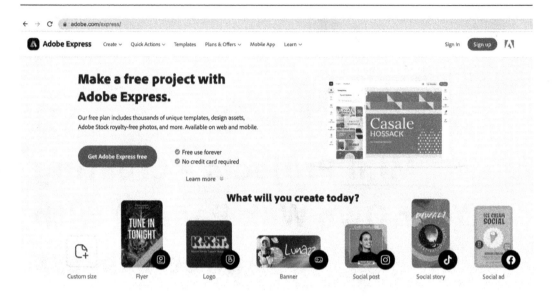

Figure 10.1 – Accessing Adobe Express via the browser

2. Once you log in, click on the + button, located on the top left of the home page, to create a new project.

Figure 10.2 – Start creating by clicking on the + button

3. Click on **Web page** to create a new project.

Figure 10.3 – Creating a new web page

4. Adobe Express will launch a new project, featuring a blank web page ready for you to begin building out your web page.

Figure 10.4 – New web page

5. To start building your web page, simply follow the call-out tab prompts located at the bottom of the page. There are three layout options for the first page, **Photo**, **Short Cover**, and **Split Layout**. For this exercise, let's opt for the **Photo** layout, but feel free to explore the other two layout options as well to enhance your practice.

    Select the **Photo** option and then proceed to either upload your own image or browse through the Adobe Stock library. I chose the **Photo** option and conducted a search on Adobe Stock using the keywords `content creator`. Click on the image of your choice and Express will automatically add this image to the web page.

Figure 10.5 – Adding an image from Adobe Stock to the first page of your web page

6. To replace the text, click on **Add a title** and replace it with your own text. I replaced the text with the name of the persona we are using for this exercise, `Sofia Lopez`. Repeat this for **Add a subtitle** and replace the text with `UGC Creator`.

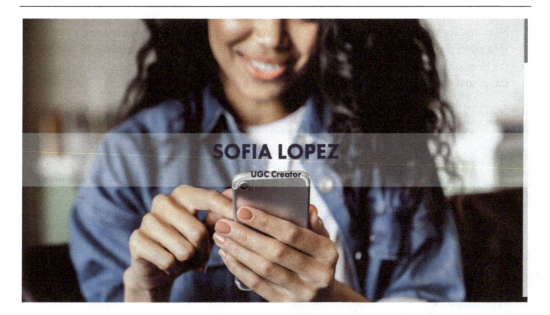

Figure 10.6 – Replacing the text with your own text

7. Scroll down and Express will display options to add **Photo**, **Text**, **Button**, **Video**, **Photo grid**, **Glideshow**, or **Split layout** components to the next page.

Figure 10.7 – Simply scroll down and Express will guide you to add content to your web page

8. Next, opt for the **Split layout** option, and Express will create a new page featuring a vertical split down the center. On one side, you can include an image, while on the other side, you have the flexibility to add various content, such as text. If you wish to interchange the layout of the image and text, simply click on the arrows located in the middle of the page.

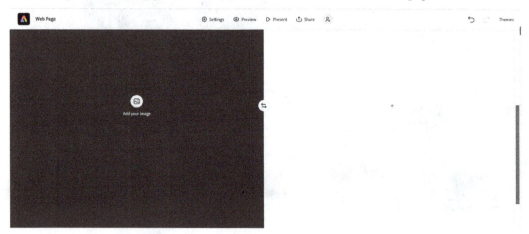

Figure 10.8 – Selecting Split layout to create a web page with two columns

9. Click on the **Add your image** option, and you can proceed to either upload your own photo or search for one in the Adobe Stock library. In this exercise, I opted to select an image from the Adobe Stock library. By clicking on the thumbnail of your chosen image, Express will automatically add the image to your page.

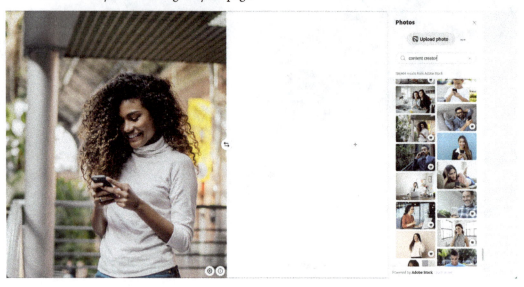

Figure 10.9 – Adding an image from Adobe Stock

10. Next, click on the + button located on the opposite side of the layout. Express will present options to add various types of content, including **Photo**, **Text**, **Button**, or **Video**.

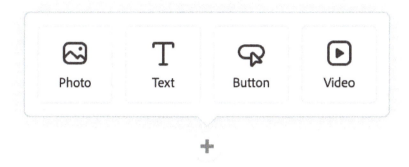

Figure 10.10 – Options to add photo, text, button, or video content

11. Select the **Text** option and input your own text. For this exercise, I included a fictional description of a person called Sofia. Feel free to incorporate your desired text based on your specific requirements for your project.

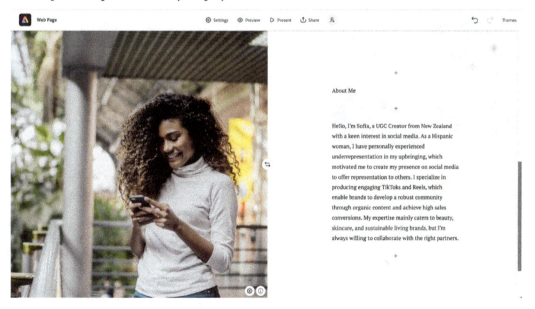

Figure 10.11 – Adding text to this page

12. After successfully adding your page content, simply click on the text you have inserted. Express will then present you with a range of text formatting options, allowing you to apply styles such as bold and italics, add bullets, change the alignment, and even include hyperlinks. This provides you with the flexibility to customize and enhance the appearance of your text, depending on your preferences.

ABOUT ME

Hello, I'm *Sofia*, a *UGC Creator* from New Zealand with a keen interest in social media. As a Hispanic woman, I have personally experienced underrepresentation in my upbringing, which

Figure 10.12 – When you click on the text, Express will display text formatting options

13. With just a few clicks, you have successfully created a beautiful split layout, serving as the second page within your Express web page.

ABOUT ME

Hello, I'm *Sofia*, a *UGC Creator* from New Zealand with a keen interest in social media. As a Hispanic woman, I have personally experienced underrepresentation in my upbringing, which motivated me to create my presence on social media to offer representation to others. I specialize in producing engaging TikToks and Reels, which enable brands to develop a robust community through organic content and achieve high sales conversions. My expertise mainly caters to beauty, skincare, and sustainable living brands, but I'm always willing to collaborate with the right partners.

Figure 10.13 – Split layout with an image on the left and text on the right

14. To continue expanding your project and incorporating additional pages, simply scroll down. Express will once again present you with a range of options to add various elements, including **Photo**, **Text**, **Button**, **Video**, **Photo grid**, **Glideshow**, or **Split layout**. This enables you to seamlessly build upon your project and create a dynamic and engaging web experience.

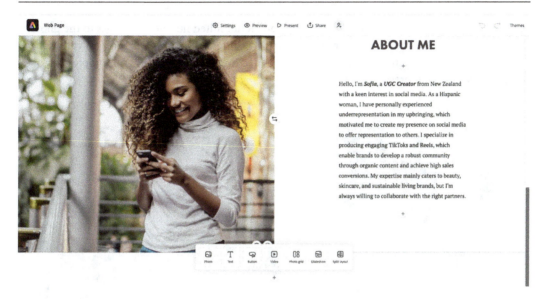

Figure 10.14 – Scrolling down to keep adding pages

15. For this exercise, we are focused on building a portfolio, so our next step involves incorporating a photo grid to showcase the recent projects completed by Sofia. Select the **Photo grid** option and Express will present you with the option to either upload your own photos or select images from Adobe Stock.

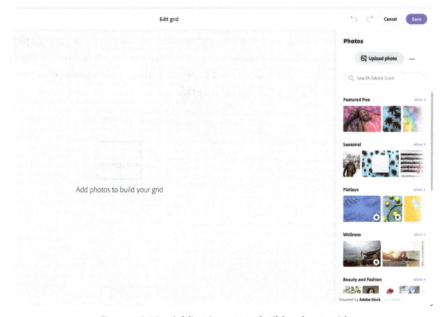

Figure 10.15 – Adding images to build a photo grid

16. With Express, you have the flexibility to upload an unlimited number of images to the photo grid. Additionally, you can easily replace, resize, reposition, and remove the images within the grid, according to your needs. Once you have added images to the photo grid, simply click on the **Save** button to preserve your changes.

Figure 10.16 – Adding images to your photo grid

17. After you have made your image selections or uploaded the desired images, Express will redirect you to the web page, where you can view your images arranged within the photo grid. To include a caption, simply click on the **Add a caption** option. This allows you to provide a description for the images displayed in the photo grid.

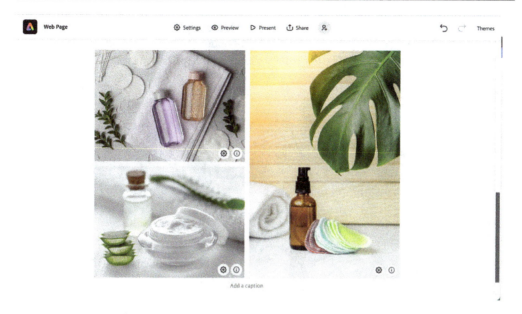

Figure 10.17 – To add a caption, click on Add a caption

18. In this exercise, I replaced the text with `Skincare UGC Photography`.

Skincare UGC Photography

Figure 10.18 – Adding a caption to your photo grid

19. To continue expanding your project and incorporating additional pages, simply scroll down. Express will once again present you with a range of options to add various elements, including **Photo**, **Text**, **Button**, **Video**, **Photo grid**, **Glideshow**, and **Split layout**.

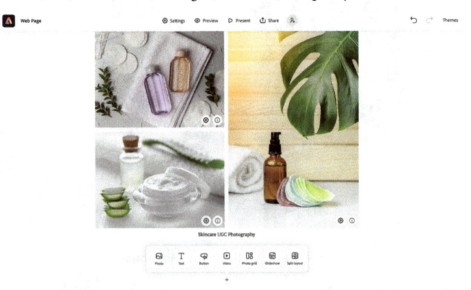

Figure 10.19 – Scrolling down to keep adding content and pages

20. Select the **Glideshow** option and, similar to *step 16*, you can proceed to upload your own images or upload images from the Adobe Stock library.

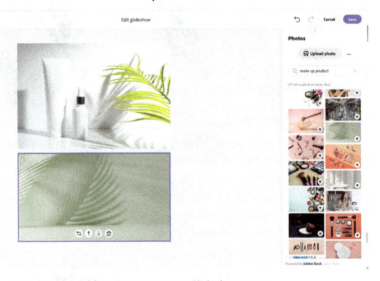

Figure 10.20 – Adding images to your Glideshow page

21. Once you click **Save**, Express will redirect you back to the web page. With the **Glideshow** page layout, Express will display a rectangular box, accompanied by a + button, allowing you to add additional content to this specific page. This feature enables you to incorporate additional elements, such as text, into your **Glideshow** page.

Figure 10.21 – A glideshow layout in Express

22. Once you click on the + button, Express will present you with a selection of options to add various elements, including **Photo**, **Text**, **Button**, or **Video**.

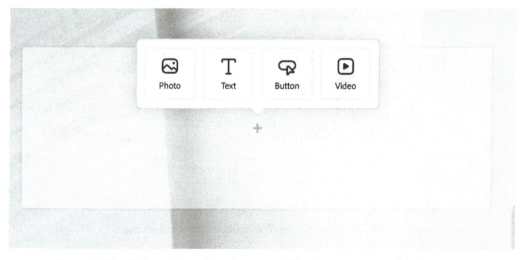

Figure 10.22 – Adding a Photo, Text, Button, or Video element to your glideshow page

23. To embed a video to your page, simply click on the **Video** option. Express offers various options to embed a video, including adding a link from popular platforms such as **YouTube**, **Vimeo**, or even directly from **Express** itself.

Figure 10.23 – Adding a video to your web page

24. Once you have added your link, click on the **Save** button. To include a title for your added content, simply click on the + button and proceed to input the desired title.

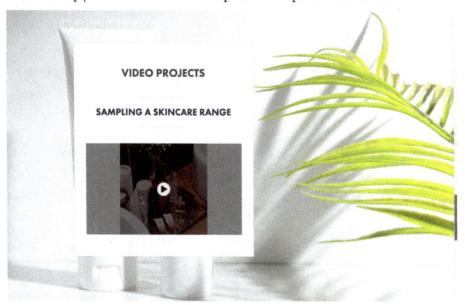

Figure 10.24 – The video has been embedded into your web page

25. As demonstrated, expanding your Express project with additional pages is a straightforward process. For instance, when building a UGC portfolio, you have the option to include a rate card, embed more videos using additional **Glideshows**, or showcase more photos from your past projects by adding more photo grids.

    The beauty of using Express lies in its flexibility, allowing you to continually build and customize your web pages according to your preferences. Whether you choose to add more content or streamline it, Express enables you to create a personalized and tailored web experience that aligns with your vision.

26. Next, click on the **Theme** button located in the top-right corner of the page. Express will present you with a collection of in-built themes that alter the overall appearance of your web page, encompassing font styles and transitions. To get a preview of each theme's visual presentation, simply select and review them one by one. Upon clicking on a theme, Express will automatically apply the chosen theme to your web page. Once you have made your selection, click on **Themes** again to close the pane and finalize your theme choice.

Figure 10.25 – Changing the theme

27. Next, click on the **Share** button at the top of the page in order to publish your web page. Click on the **Publish and share link** option to generate a shareable link.

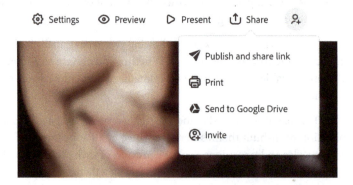

Figure 10.26 – Sharing your web page

28. Complete the required fields by providing a title for your web page, selecting a suitable category, and deciding whether you wish to publish your name as the author of this web page. Once you have filled in these details, click on the **Publish** button.

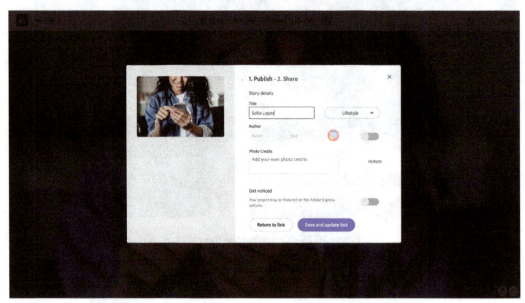

Figure 10.27 – Completing the fields for publishing

Congratulations on successfully creating your first Express web page! You can now easily view your web page by accessing it through the provided hyperlink. Take a moment to celebrate your accomplishment!

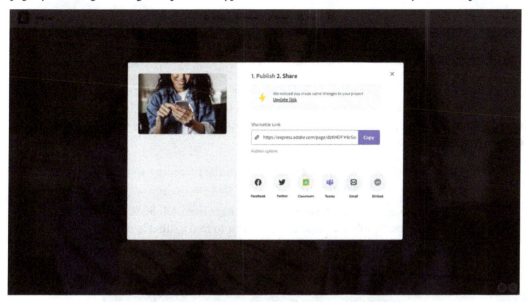

Figure 10.28 – Express will host your web page on Adobe Express

Well done! You have completed this exercise successfully. The skills you have acquired in this exercise can be applied to create your own portfolio, resume, or any other project you have in mind.

Figure 10.29- To view the finished web page, scan the QR code or visit https://bit.ly/sofialopezugc

Throughout this section, you have explored a comprehensive step-by-step guide walkthrough, guiding you through the process of creating a web page. This section highlighted the versatility of web pages built in Express, showcasing the ability to create a web page for a portfolio.

In the upcoming section, we will explore another exercise, guiding you through the process of creating a marketing splash page specifically designed for an e-commerce business.

Designing a marketing splash page for an e-commerce business

In this exercise, we will focus on creating a marketing splash page tailored to an e-commerce business. For this exercise, we will explore the example of a fictitious indoor plant shop. Similar to the previous chapter, this exercise will provide you with valuable hands-on experience as you follow a detailed step-by-step guide to build out this web page:

1. To get started, simply follow *steps 1-4* as outlined in the preceding section. These steps will serve as a foundation to kickstart your progress in this exercise.

2. Click on the **Photo** option and proceed to add an image from Adobe Stock. For this exercise, conduct a search specifically for `indoor plants` to find a suitable image that aligns with the theme.

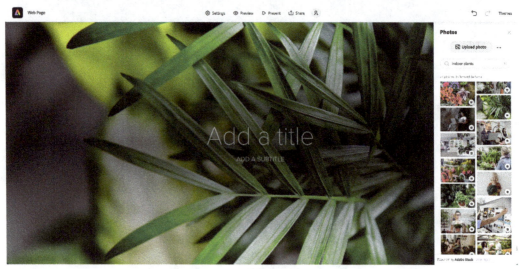

Figure 10.30 – Adding an image to the front page

4. Replace the text with `The Kiwi Jungle Oasis`. To continue expanding your project and incorporating additional pages, simply scroll down to add more content.

Figure 10.31 – Adding a title to your web page

5. I opted for the **Glideshow** option, and here you will be presented with the choice to either upload your own images or access images from Adobe Stock. In this instance, select two photos from Adobe Stock and proceed by clicking the **Save** button.

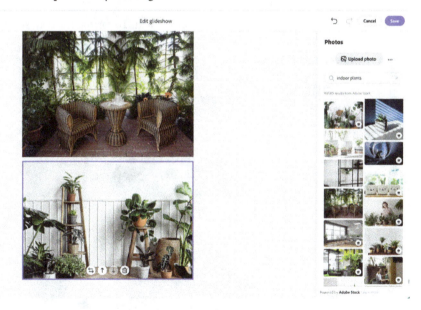

Figure 10.32 – Adding photos from Adobe Stock

6. After clicking on **Save**, Express will redirect you back to the web page. With the **Glideshow** page layout, you will notice a rectangular box accompanied by a + button, allowing you to add additional content, such as a **Photo**, **Text**, **Button**, or **Video** option.

Figure 10.33 – Adding additional content by clicking on the + button

7. Select the **Text** option and input your desired text. To format the text, simply click on the text itself, and Express will display various text formatting options.

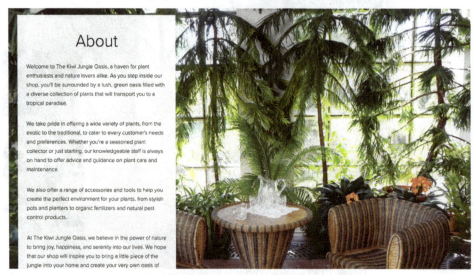

Figure 10.34 – Adding text to your glideshow page

8. Once you have included text on your **Glideshow** page, continue by scrolling down to incorporate additional content. Express will then present the call-out tab prompts located at the bottom of the page, featuring a variety of content elements to choose from.

Figure 10.35 – Scrolling to add more content and pages

9. Opt for the **Split layout** option, and Express will generate a new page featuring a vertical split down the center. On one side, you can easily include an image, while on the other side, you have the flexibility to add various content such as text. If you wish to interchange the layout of the image and text, simply click on the arrows located in the middle of the page.

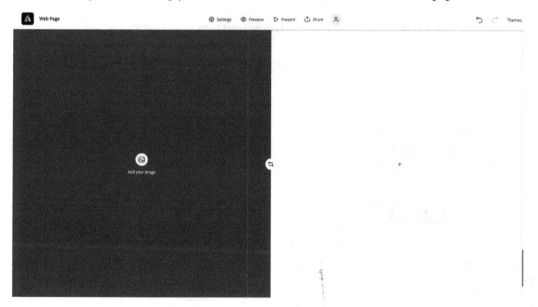

Figure 10.36 – Creating a split layout

10. When you select the **Add your image** option, you will have the choice to either upload your own photo or explore the Adobe Stock library. Then, proceed to the opposite side of the split layout and select the + button and select the **Text** option. Feel free to input your own text based on your project requirements.

Caring for your plants

Watering indoor plants is a crucial part of plant care, as it helps to keep them healthy and thriving. Here are some steps to follow when watering your indoor plants:

1. Check the soil moisture level by inserting your finger about an inch into the soil. If it feels dry, it's time to water.
2. Slowly pour room temperature water onto the soil until it starts to drain out of the bottom of the pot.
3. Avoid getting water on the leaves.
4. Water indoor plants about once a week, adjusting as needed for the type of plant and the environment.

By following these simple steps, you can keep your indoor plants healthy and happy.

Figure 10.37 – Adding an image and adding text to the Split layout page

11.   To expand the content of your web page, simply scroll down and continue adding more elements. Click on the **Photo** option and proceed to include a photo from Adobe Stock.

Figure 10.38 – Adding a photo to the next page

12. For this instance, I searched for `monstera` in Adobe Stock.

Figure 10.39 – Adding an image from Adobe Stock

A. After adding your chosen image, simply click on the image itself on the web page. Express will then present you with a range of image formatting options to choose from, such as **Inline**, **Fill screen**, **Window**, **Full width**, **Move**, **Focal point**, **Replace**, and **Delete**. These versatile options enable you to select and customize the display style of your image on the page according to your preferences and requirements.

Figure 10.40 – Image formatting options

13. For this instance, I opted for the **Fill screen** option to display the image in full-screen mode. If you select this display option, the image occupies the entire screen.

Figure 10.41 – Fill screen image option

14. Proceed with adding a caption by This button isn't visible in the screenshot.

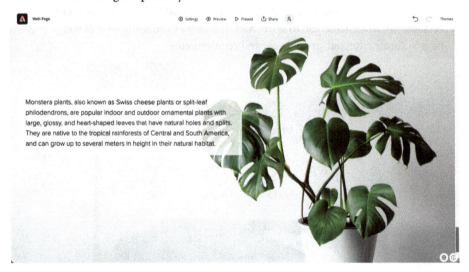

Figure 10.42 – Adding a caption to this page

15. Similar to the previous exercise within this chapter, expanding your Express project is an easy process. You have the flexibility to incorporate additional content, including supplementary product images, a FAQ page, and detailed information about the physical store for this specific project. This allows you to further enrich and personalize your web page to provide a comprehensive and engaging experience to your audience.

16. After incorporating all the desired content for this web page, simply proceed to follow *steps 26-29* as outlined in the previous exercise. These steps encompass modifying the theme of your web page and finalizing the publishing process. By following these steps, you can customize the overall appearance of your web page and then publish it to make it accessible to your intended audience.

The advantage of using Express is that you can continue to build and personalize your web page by adding more content according to your preference.

Congratulations on completing this exercise successfully! You can apply the skills you've acquired to develop your own e-commerce or marketing page.

Figure 10.43 – To view the web page created for this exercise, scan
the QR code or visit https://bit.ly/kiwijungleoasis

Summary

Throughout this chapter, you gained valuable knowledge on creating web pages by engaging in two practical exercises. I hope that you have successfully completed these exercises and are now able to showcase web pages that you can take pride in. This chapter emphasized the versatility of web pages, illustrating their ability to serve various purposes, including marketing splash pages, presentations, portfolios, resumes, registration pages, newsletters, advertisements, business proposals, and product catalogs.

Web pages provide individuals and businesses with a cost-effective and efficient tool to establish an online presence, build their brand identity, engage with customers or an audience, and obtain valuable insights into the preferences and needs of their target audience.

With Adobe Express, it is no longer necessary to hire a professional or learn coding to build a web page. This chapter demonstrated the ease and accessibility of creating professional and visually appealing web pages with Adobe Express.

Part 4 – Create a Video with Adobe Express

Welcome to the part of the book that focuses on creating, editing, and scheduling videos with Adobe Express. Throughout this section, you will learn how to create polished videos using a variety of features, such as adding videos, images, icons, text, and music. You will also learn how to adjust audio, record voice-overs, and share your final videos.

Additionally, you will explore video Quick Actions in Express, powered by Adobe Premiere Pro, which enables you to effortlessly enhance your videos. This feature provides you with the ability to resize, convert to GIFs, crop, adjust speed, convert to MP4, and much more, all in just a few minutes.

Finally, you'll learn how to automate the process of scheduling and posting your content on multiple social media platforms, using the Content Scheduler in Express. With this knowledge, you can plan and batch-process your content ahead of time, building a social media strategy that works for you. By the end of this section, you will have all the tools you need to create stunning videos and effortlessly share them with the world.

This part has the following chapters:

- *Chapter 11, Creating and Editing Videos*
- *Chapter 12, Polishing Videos Using Quick Actions*
- *Chapter 13, Scheduling Content in Adobe Express*

11

Creating and Editing Videos

In this chapter, you will learn how to create and edit videos with Express. You can quickly and easily create videos from the browser or a mobile device; there is no need to download any software. This chapter will show you how to use Express for video editing. You will learn how to add videos and images to create a polished video. You will also learn how to record audio and add icons and music to your videos.

We will cover the following topics in this chapter:

- How to create a video from scratch and how to upload videos
- How to adjust audio – adjusting the volume, recording a voiceover, and adding a soundtrack
- How to add text, images, and icons to videos
- How to download and share videos

By the end of this chapter, you will be able to create and edit videos quickly. Express is a great tool for making and editing videos. It is an easy-to-use tool that is easy to learn, whether you are a complete beginner or a professional.

Creating and uploading videos from scratch

To get started, follow these steps on the browser:

1. Navigate to the home page on your desktop: `https://express.adobe.com/`.

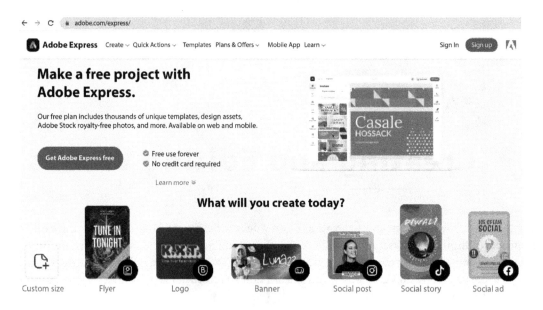

Figure 11.1 – Access Adobe Express via the browser

2.   Once you are logged in, click on the + button to start a new project.

Figure 11.2 – Start creating by clicking on the + button

3. Next, click on **Video**.

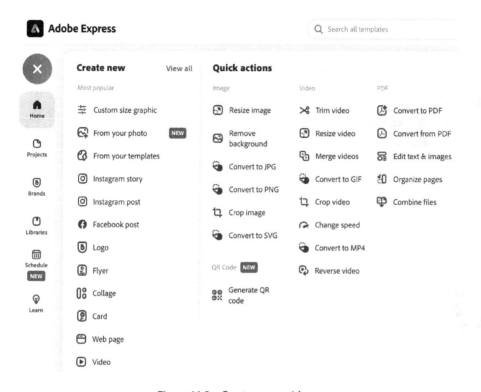

Figure 11.3 – Create a new video

4. Upon reaching this page in Express, you will be presented with several options. You can either choose to utilize their template, which we will cover shortly, by inputting a title for your video, or you can simply click on the **Skip** button. Opting to click on **Skip** allows you to create a video from scratch, starting with a blank canvas.

Figure 11.4 – Two options – start from a template or scratch

5.   Once you click on **Skip**, Express will present you with a selection of templates to choose from. Alternatively, you can click on **Start from scratch** to begin creating your video without utilizing any pre-designed templates.

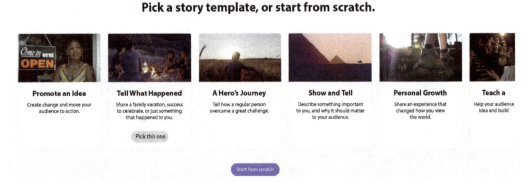

Figure 11.5 – Choose a template or create from scratch

Express will take you to a new project.

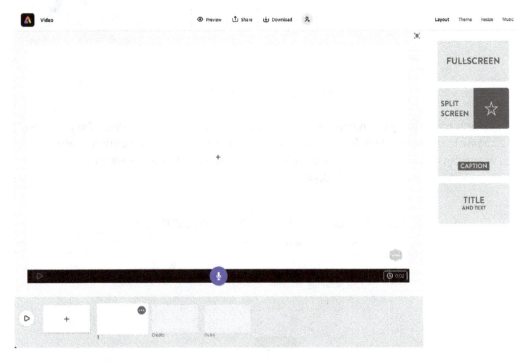

Figure 11.6 – Blank project

6. On the right-hand side, locate the **Layout** option and click on it. From the available choices, select **Fullscreen**. Notice on the preview screen there is a + icon, indicating the option to add either a video or photo. Proceed by clicking on the **Video** option.

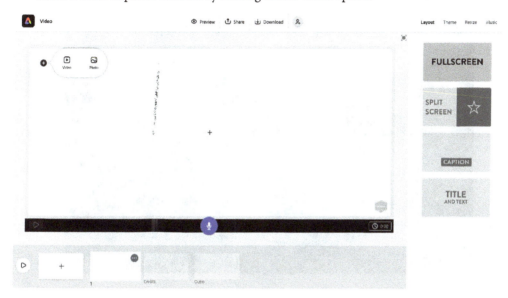

Figure 11.7 – Fullscreen layout

7. Locate the video file you wish to upload and click **Open** to proceed. Express will then provide a preview of your selected video, allowing you to make further adjustments if desired. You can trim your video by dragging the circles at the beginning or end of your clip. Once you have trimmed your video, click **Save**.

Figure 11.8 – Add and trim your video

8. Express will automatically incorporate the uploaded video into your timeline. In the bottom-right corner, you will find the timecode indicating the current duration of the video (e.g., **0.14**), while the play button is located at the bottom left.

Figure 11.9 – Video preview

9. In the top-right corner, you will find three options: **Zoom, Trim, and Clip Volume**. Let's explore each option to understand their functionalities.

Figure 11.10 – Video options

A. To adjust the zoom level of the video or image, click on the **Zoom** option represented by a magnifying glass icon. Clicking the + button will zoom in while clicking the – button will zoom out. To navigate back to the main menu, click on the ellipsis menu (three horizontal dots), also known as the meatballs menu.

Figure 11.11 – Video has been zoomed in

B. To trim your video, click on the Trim option, represented by a pair of scissors icon, as per *Figure 11.10*. This action will open a separate window where you can perform the trimming. Simply click and drag the circles located at the start or end of the video to trim the desired portions. Once you have trimmed your clip, click on the **Save** button to apply the changes.

Figure 11.12 – Trim option

C.   In the upcoming section, we will explore the **Clip Volume** option, depicted in *Figure 11.10*, to further understand its functionality and usage.

10. If you have navigated away from it while exploring the preceding step, simply return to the main project and proceed to *step 9*.

Underneath the video preview, you will find the timeline displaying video frames. To add a new frame, simply click on the + icon located on the thumbnail.

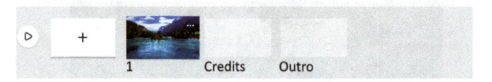

Figure 11.13 – Add frames to the timeline

11. When you add a new frame, Express will display a new blank frame in the video preview. Within this frame, you will find options to include various content types, such as video, text, photo, or icon.

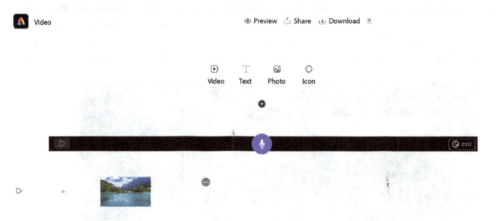

Figure 11.14 – Add different content to the new frame

12. Select the **Video** option to upload an additional video.

13. Similar to *step 8*, you have the option to trim your video or keep it as is, then click on the **Save** button.

Figure 11.15 – Add and trim your video

14. Notice that you now have two frames in your timeline. To switch between the videos, click on the thumbnail corresponding to the desired video and it will be displayed in the preview window.

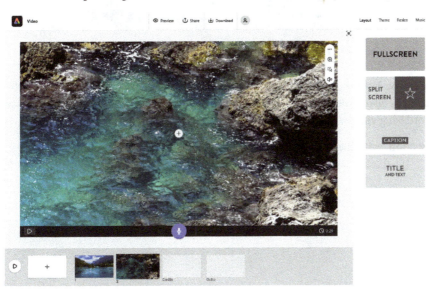

Figure 11.16 – There are now two videos in your timeline

15. To rearrange the videos, simply click and drag the thumbnail to your preferred position, allowing you to reorder the video frames.

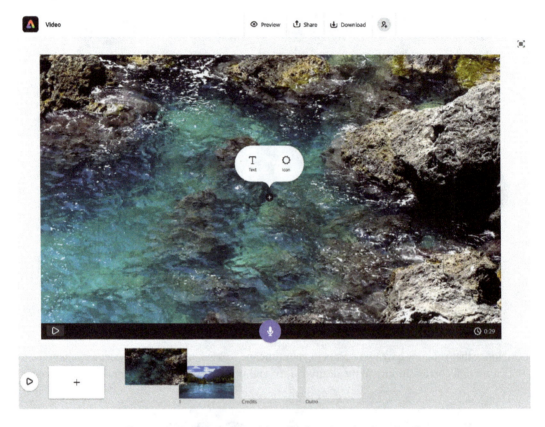

Figure 11.17 – Reorder the videos by dragging the thumbnails

In this section, you have acquired the skills to create a new video project and upload videos to the timeline. You have also gained knowledge on trimming videos prior to adding them to the timeline, as well as reordering videos.

In the next section, we will explore the process of adjusting various audio components.

How to adjust audio – adjusting the volume, recording a voiceover, and adding a soundtrack

In this section, we will explore the techniques for adjusting audio components in your videos. You will discover how to modify the volume of videos, record voiceovers, and incorporate soundtracks to enhance your video projects:

1. Click on the pencil icon in the top-right corner of the video preview.

Figure 11.18 – Click on the pencil icon

Express will display several editing options for your video.

Figure 11.19 – Video editing options

2. Clicking on the speaker icon (the last icon) will reveal three volume options: **Loud**, **Soft**, and **Mute**. Select your preferred volume to adjust the volume of videos.

Figure 11.20 – Adjust the volume

3. To record a voiceover, click on the microphone icon in the purple circle in the middle of your video preview.

Figure 11.21 – Record a voiceover

4. To record a voiceover, simply press and hold down the microphone icon button located within the purple circle positioned at the center of your video preview. To stop recording, release the button.

Figure 11.22 – Record a voiceover by holding down the microphone icon button

5. To add music to your video project, navigate to the **Music** tab located in the top-right corner of the canvas.

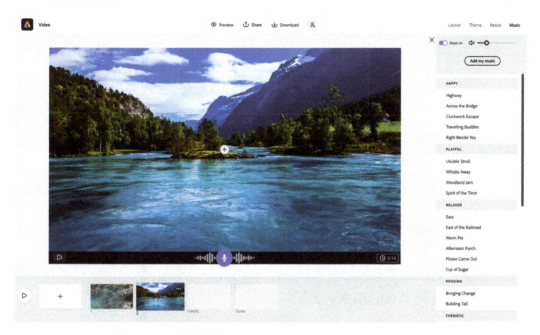

Figure 11.23 – Add a soundtrack

6. Express offers a selection of royalty-free music for you to utilize in your project. To preview a specific music track, simply click on the play button located to the left of the track's title.

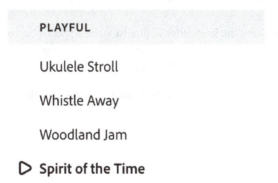

Figure 11.24 – Preview the music

7. To add your own music, click on the **Add my music** button and proceed to upload your own soundtrack.

Figure 11.25 – Upload your own music

8. By clicking on the track, Express will seamlessly integrate the chosen soundtrack into your video. You will see a checkmark displayed next to the music, indicating its successful addition.

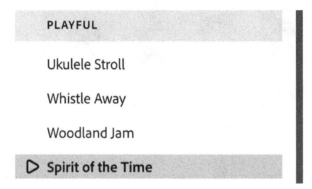

Figure 11.26 – Click to add music to your video

In this section, you have acquired the knowledge to fine-tune the audio aspects of your videos, including volume adjustment, voiceover recording, and adding music. Express empowers users of all skill levels to create professional videos with its powerful features.

In the next section, we will delve into the process of incorporating text, images, and icons into your videos, further expanding your creative possibilities.

How to add text, images, and icons to videos

In this section, we will explore how to enhance your videos by incorporating text, images, and icons. These elements can help you create more visually appealing and engaging videos. You will learn how to precisely position and time these elements, and you will receive guidance on changing the video's theme to create more dynamic and visually appealing content. To get started, let's continue working on your project by following the steps outlined here:

1. Click on the + button within the video preview.

Figure 11.27 – Click on the + button

2. When you click on the + button, Express will present you with the option of adding two different types of content: **Text** or **Icon**.

Figure 11.28 – Add text or an icon

3. Select the **Text** option and proceed to input a title, which will be positioned on top of your video.

Figure 11.29 – Add text

4. To include a subtitle, select the **TITLE AND TEXT** layout option from the **Layout** tab.

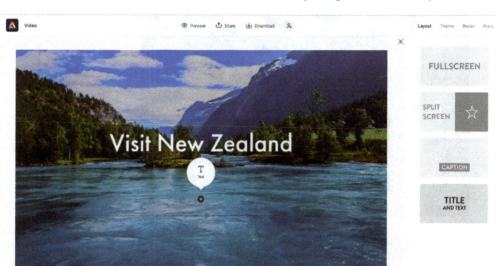

Figure 11.30 – Add a subtitle

5. To change the font size, simply click on the text and use the **T-** button to decrease the font size or the **T+** button to increase it.

Figure 11.31 – Change the font size

6. Currently, it is not possible to change the font in Express (as of the time of writing this book). However, you can change the overall theme, which will also alter the font. Simply click on the **Theme** option and browse through the available themes provided by Express. If you have created a brand within Express, you will also find your branded themes listed (as shown in the following example of my fictitious brand, **Delicious Treats**).

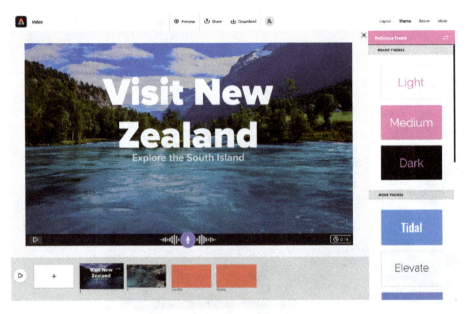

Figure 11.32 – Change the theme

7.   To include an icon in your video, click on the + button, followed by selecting the **Icon** option.

Figure 11.33 – Add an icon

8. Once you click on the icon, Express will display the icon search bar on the right side. You can enter a keyword, such as New Zealand, to explore and browse relevant icons.

Figure 11.34 – Search for an icon

9. Click on the desired icon and Express will position it over your video.

Figure 11.35 – Add the icon to your video

10. To adjust the size of your video, click on the **Resize** option and choose between the **WIDESCREEN** and **SQUARE** aspect ratios.

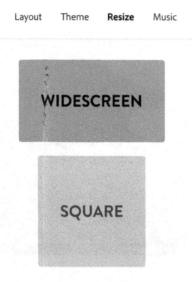

Figure 11.36 – Resize your video

In this section, you have acquired the skills to incorporate text and icons into your videos effectively. You have also gained knowledge on changing the theme, exploring a vast collection of icons, and resizing your video.

In the next section, you will discover the process of downloading and sharing your videos, ensuring your projects reach their intended audience.

Downloading and sharing videos

In this section, you will be guided through the process of downloading and sharing your videos. Now that you have successfully created a video, let's proceed with the following steps to learn how to export your video:

1. To download your video, click on the **Download** button.

Figure 11.37 – Download

Once you have clicked on the **Download** button, Express will prepare the video and download it.

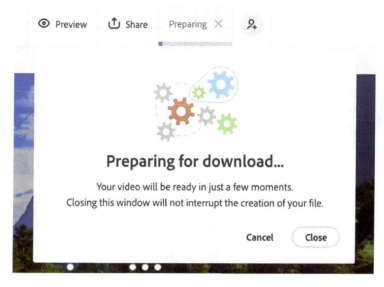

Figure 11.38 – Preparing for download

2. If you wish to generate a link for your video, you can select the **Publish** option. To begin the process, click on the **Share** button.

3. Next, simply proceed by clicking on the **Publish** button.

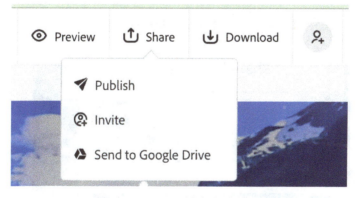

Figure 11.39 – Publish the video

4. Give your video a title and decide whether you wish to display your name alongside it. Then, simply click on the **Create link** button.

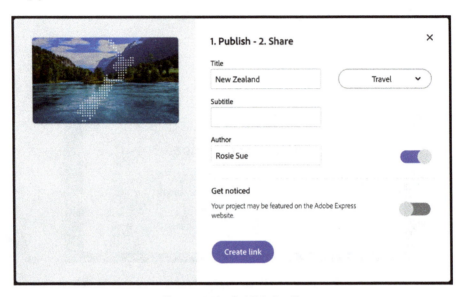

Figure 11.40 – Publish details

5. Once the link has been generated by Express, you will receive a URL that you can use to access your video.

Figure 11.41 – Express has created a URL link to your video

6. You now have the ability to share a public link to your video, which is hosted on Adobe Express. By sharing this link, anyone who has access to it can view your video.

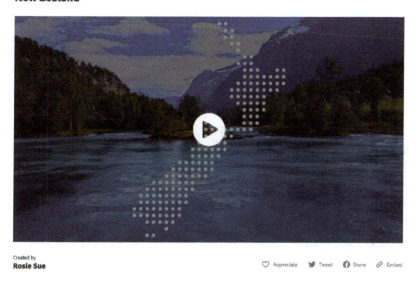

Figure 11.42 – You can now share this link to your video

Figure 11.43 – Scan this QR code to view the video created in this chapter

In this section, you have acquired the knowledge of downloading and sharing your videos. This newfound ability allows you to conveniently download your videos and easily distribute them across various platforms. This skill is valuable for content creators and individuals seeking to create videos with ease. With your polished videos at hand, you can now expand your reach and establish great engagement with your audience.

Summary

In this chapter, you have acquired the knowledge and skills to create and edit videos using Express. With this newfound ability, you can confidently produce professional-grade videos in a quick and efficient manner. This skill will enable you to edit video content for various purposes, including your business, personal, and professional brand. As video content continues to dominate the digital landscape, video content is king. Social media platforms are placing increased priority on video content and your acquired skills will keep you ahead of the curve.

The upcoming chapter will introduce you to video quick actions, powered by Adobe's core video editing app, Premiere Pro. This feature offers a comprehensive set of powerful tools that enable users to efficiently edit their videos. By exploring video quick actions, you will further enhance your video editing capabilities and expand your creative possibilities.

12

Polishing Videos Using Quick Actions

In this chapter, you will learn how to use video Quick Actions, a set of powerful features integrated from Adobe's renowned core video editing app, Premiere Pro. As the leaders in the video editing space, these powerful features enable you to quickly and easily polish your videos in a matter of minutes. You will discover how to resize videos to various aspect ratios, including popular dimensions for platforms such as Instagram, Facebook, and YouTube. Additionally, you will gain insights on converting videos to GIFs, cropping videos, adjusting their speed, converting them into MP4 format, and even reversing their playback.

We will cover the following topics in this chapter:

- How to reformat videos (trim, resize, merge, and speed up/down)
- How to convert videos to GIFs
- How to convert into MP4

By the end of this chapter, you will have gained the skills to efficiently refine your videos using Express' quick video actions. These powerful tools will enable you to swiftly make precise adjustments to your videos, enhancing their quality and impact. With the knowledge you will gain, you will be able to fine-tune your videos with ease, achieving the desired results in no time.

How to reformat videos (trim, resize, merge, and speed up/down)

To begin with, let's explore the process of trimming and resizing a video.

To get started, follow the steps outlined here from the browser:

1. Navigate to the Adobe Express home page, `https://express.adobe.com/`, in your browser:

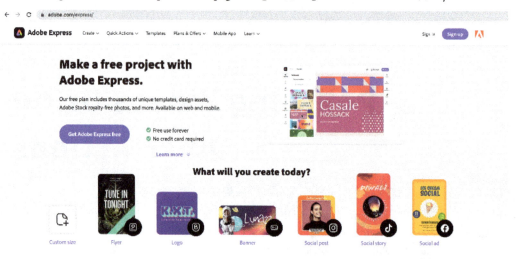

Figure 12.1 – Access Adobe Express via the browser

2. After logging in, click on the + button in the top-left corner.

Figure 12.2 – Start creating by clicking on the + button

3.   Next, click on **Trim video**.

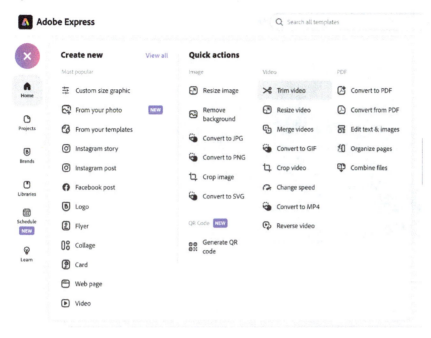

Figure 12.3 – Trim video

4.   Click on **Browse on your device** to upload a video. Alternatively, you can use the sample video.

Figure 12.4 – Trim video

5.   Once you upload your video, Express will display your video, along with the timeline at the bottom.

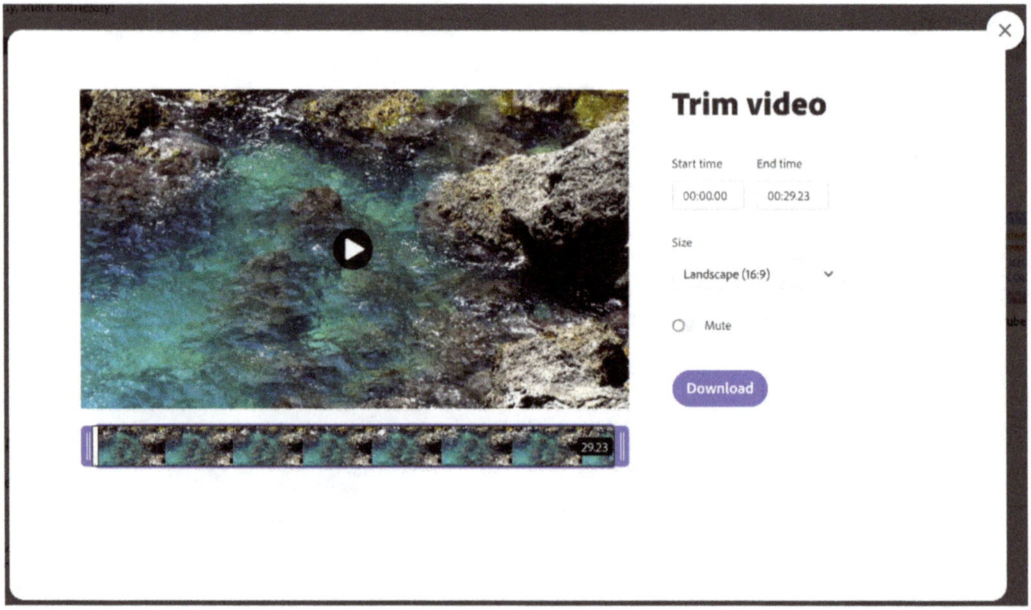

Figure 12.5 – Trim video options

6.   Simply click and drag the purple rectangles on either side of the timeline to trim the beginning or end of your clip.

Figure 12.6 – Drag the sides of the timeline to trim your video

7.   Alternatively, you can input a precise time code in the **Start time** and **End time** fields located to the right of the video.

You can also change the aspect ratio of your video by clicking on the drop-down box under the **Size** option. You can change the aspect ratio to the following sizes: **Landscape (16:9)**, **Portrait (9.16)**, or **Square (1:1)**.

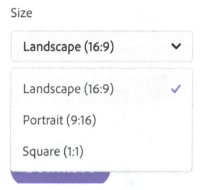

Figure 12.7 – Choices to change the aspect ratio of your video

8.   You can also mute the video by toggling the **Mute** option.

9.   Once you have trimmed your video, simply click on **Download** to download a copy of your trimmed video.

Figure 12.8 – Express will show you the progress

10. After the video has been processed by Express, you will see the following message, which provides you with the option to try out Premiere Pro or start creating.

Figure 12.9 – Successful download message

11. Next, let's explore the process of merging videos. Please follow the instructions outlined in *steps 1* and *2* of this section.

12. Click on **Merge videos**.

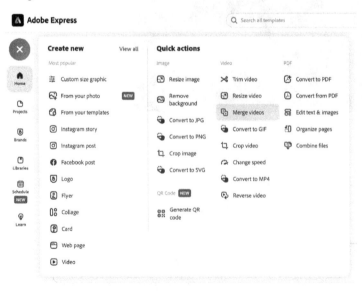

Figure 12.10 – Click on Merge videos

13. Express will display the window to upload your video. You can either drag and drop a video or click on **Browse on your device**.

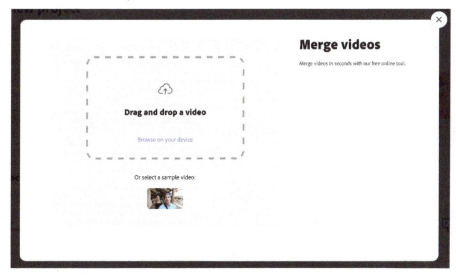

Figure 12.11 – Upload a video

14. Upload the videos you want to merge. Express will display thumbnails of all the uploaded videos, depending on the number of videos you have uploaded. You can click on the + button to add more videos if needed.

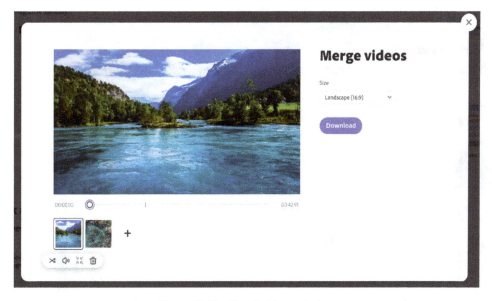

Figure 12.12 – Merge videos window

15. Click on **Download** to download your video compilation.

16. Next, we will explore the process of speeding up and slowing down your videos using quick actions.

17. Please follow the instructions outlined in *steps 1* and *2* of this section.

18. Click on **Change speed** in the **Video** category under **Quick actions**.

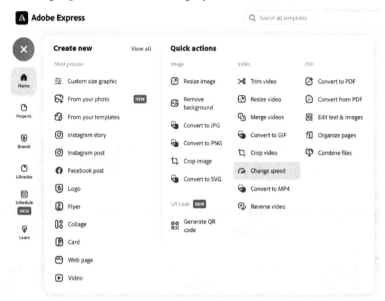

Figure 12.13 – Change speed

19. Upload your video.

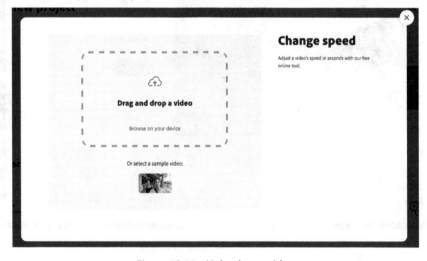

Figure 12.14 – Upload your video

20. Express will display several options to change the speed of your video. These are **Super slow (25%)**, **Slow (50%)**, **Normal (100%)**, **Fast (150%)**, and **Super Fast (200%)**.

Figure 12.15 – Speed options

21. Simply click on your desired speed and click **Download**.

In this section, you acquired valuable video editing techniques such as trimming, resizing, merging, and speeding up videos. These video editing techniques have become increasingly easy with Express, making video editing more accessible than ever before.

In the next section, we will explore the process of converting videos into GIFs.

Converting videos into GIFs

In this section, we will learn the process of converting videos into GIFs:

1. To get started, please refer to *steps 1* and *2* in the previous section. Following that, click on the **Convert to GIF** option.

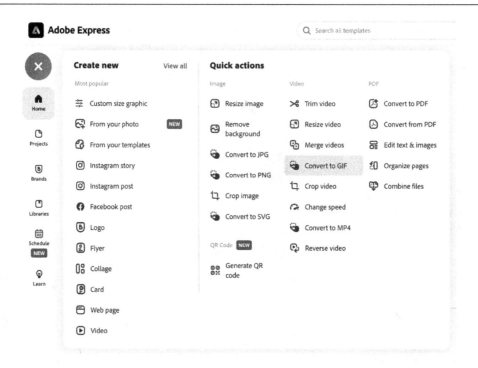

Figure 12.16 – Convert to GIF

2.   Upload your video.

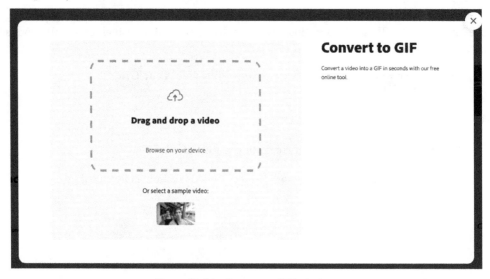

Figure 12.17 – Convert to GIF window

3. Choose the specific segment of your video that you wish to convert into a GIF, and then select the desired size and aspect ratio.

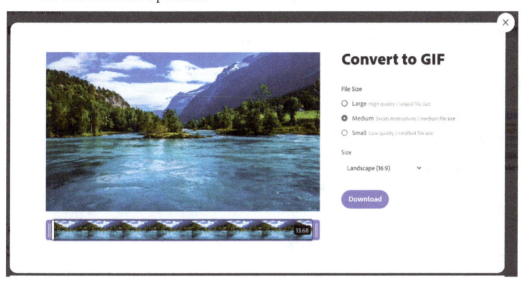

Figure 12.18 – Select a portion of your video

4. Click on **Download**, and Express will initiate the download process for your GIF, providing you with a copy of your converted GIF file.

Figure 12.19 – Copy of the GIF

In this section, you acquired the skill of converting your video into a GIF. By following the straightforward process of uploading your video and initiating the conversion, you can easily obtain a GIF copy of your video.

In the next section, we will explore the process of converting videos into MP4 format.

How to convert into MP4

In this section, we will explore the process of converting videos into MP4 format.

To get started, follow these steps in the browser:

1. To get started, please refer to *steps 1* and *2* from the *How to reformat videos (trim, resize, merge, and speed up/down)* Following that, click on the **Convert to MP4** option.

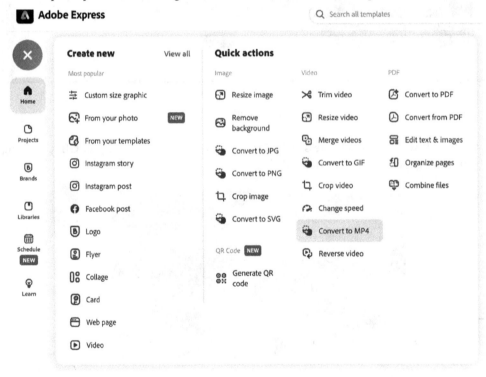

Figure 12.20 – Convert to MP4

2. Upload your video.

Figure 12.21 – Upload your video

3. Express will upload the media, and once it's ready, you will see the **Download** button.

Figure 12.22 – Download your file

4. You now have a copy of the MP4 file.

Lake_-_52849_AdobeExpress.mp4
MPEG-4 movie - 9.8 MB

Information	Show More
Created	Today, 7:56 PM
Modified	Today, 7:56 PM
Dimensions	1280×720
Duration	00:14

Tags
Add Tags...

Figure 12.23 – MP4 file

In this section, you acquired the skill of converting any video file format into an MP4 file.

Summary

In this chapter, you gained valuable insights into the convenient and efficient video editing capabilities offered by Express and its video quick actions tools. You can now easily merge, trim, modify the size of, and adjust the speed of your videos. Additionally, you discovered the simplicity of converting videos into GIFs and converting any file format into an MP4 file.

In the next chapter, we will explore the process of utilizing the Content Scheduler in Express to plan and publish your content across multiple social media platforms.

13

Scheduling Content in Adobe Express

In this chapter, you will learn how to use the **Content Scheduler** in Adobe Express to plan and publish your content on multiple social media platforms. The Content Scheduler will allow you to automate the process of scheduling and posting your content, with the ability to batch-process on multiple platforms simultaneously.

The Content Scheduler allows you to batch-create content and plan ahead, which enables you to build a social media strategy and plan your content in advance.

With the growing number of social media platforms, it can be daunting to keep up with the demand. However, by utilizing the Content Scheduler in Express, you can plan ahead of time. Whether you are in marketing and need to create several social media campaigns each month, or you have a social media presence for your personal brand, using the Content Scheduler allows you to automate these time-consuming processes.

We will cover the following topics in this chapter:

- How to schedule your content
- How to preview your content before a post goes live

By the end of this chapter, you will be able to schedule your content with ease. There will be no need to manually upload your content to all the various social media channels; instead, you will now have the option to automate this process.

How to schedule your content

To get started, follow these steps on the browser:

1. Navigate to the home page on your desktop: `https://express.adobe.com/`.

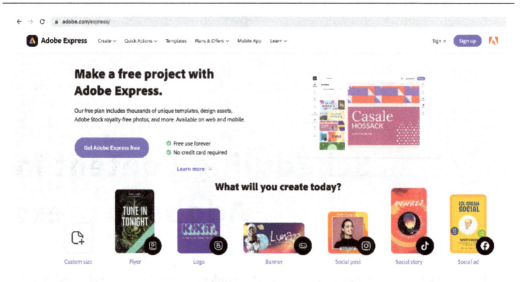

Figure 13.1 – Accessing Adobe Express via the browser

2.   Navigate to the **Schedule** option, located on the left side of the page.

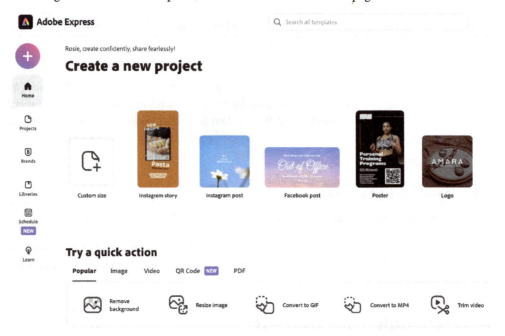

Figure 13.2 – Clicking on Schedule

After clicking on the **Schedule** button, Express will display two options:

- You can connect to your social media channels, or you can choose to skip this step for now. For this exercise, let's opt for the **Skip - connect later** option.

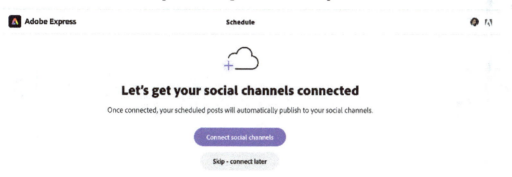

Figure 13.3 – Connecting to social channels

- Express will present a calendar interface where you can manage scheduling. On this page, you can toggle between two view options – **Week** or **Month**. Additionally, you can also navigate to a specific week or month as needed. To add a new entry to the calendar, simply click on **Add new**.

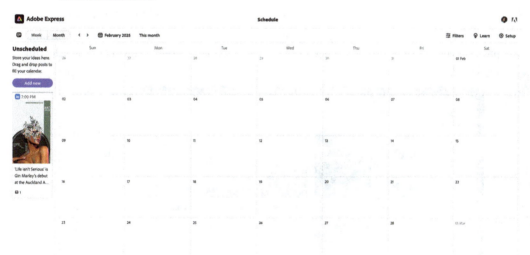

Figure 13.4 – The calendar

3.   When you click on **Add new**, Express will display a window to create a new post.

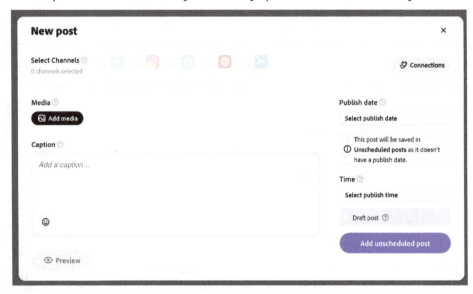

Figure 13.5 – Creating a new post

4.   On this interface, proceed to add your content by clicking on the **Add media** option.

5.   Express offers two methods to upload your content. You have the option to browse through projects you have created within Express, or you can upload your own content by selecting **My Device**.

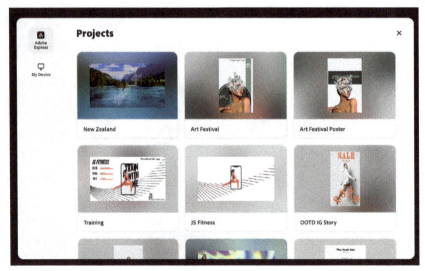

Figure 13.6 – Uploading your content

6. Click on one of your projects, and you will have the option to select multiple pieces of content. For this example, we will choose only one project, and then click on **Select 1**.

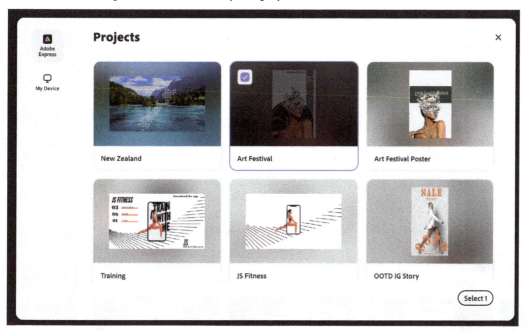

Figure 13.7 – Selecting content

7. After Express has finished loading your file, proceed by selecting it, and then click on **Add to post**.

Figure 13.8 – Add to post

8.   Express will proceed to load your media into the new post.

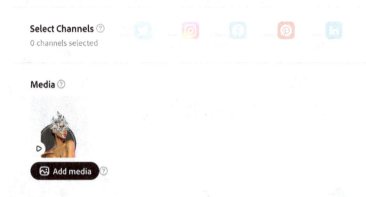

Figure 13.9 – Express will add your media to the new post

9.   Following that, proceed to write a caption for your post. Additionally, you have the option to include emojis in your post if desired.

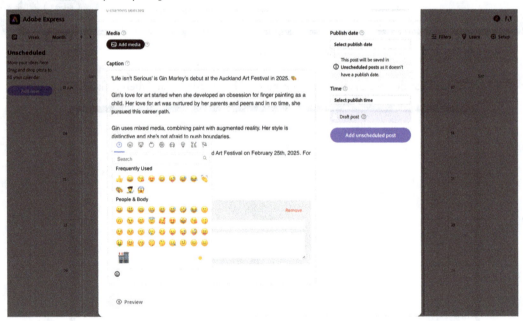

Figure 13.10 – Adding a caption

10. In the top right, select your publish date.

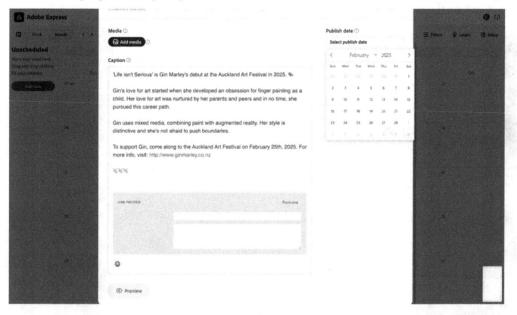

Figure 13.11 – Publish date

11. Underneath the publish date, select the preferred scheduled time for your post.

Figure 13.12 – Scheduling a posting time

12. Next, select the social media platforms where you want to publish your scheduled content. Click on the **Connections** button to proceed.

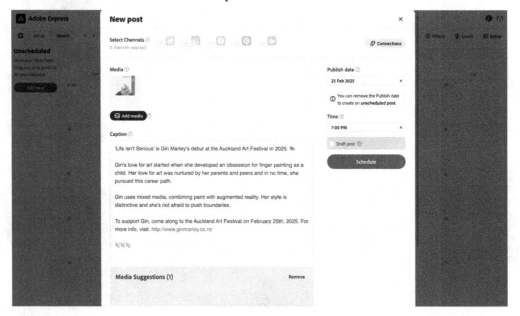

Figure 13.13 – Connecting to your social media accounts

13. Click on **Connect** to connect to your social media accounts.

Figure 13.14 – Connecting to the various social media platforms

When you connect your social media accounts, Express requests your permission to access the following information.

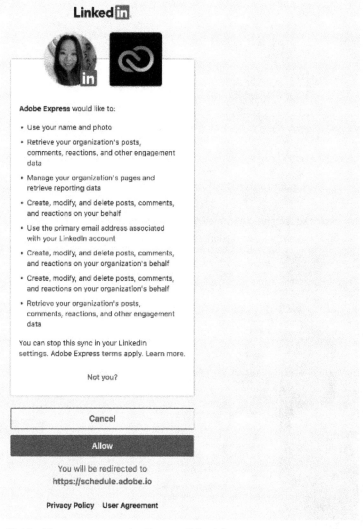

Figure 13.15 – The terms to use the Content Scheduler

14. Once you grant access to Express, you will receive a notification indicating a successful connection.

Channel successfully connected

Return to Schedule

Figure 13.16 – The display message once your account is connected to Express

15. After successfully connecting your social media accounts, you can select your desired channel in the new post, and then click on **Schedule**.

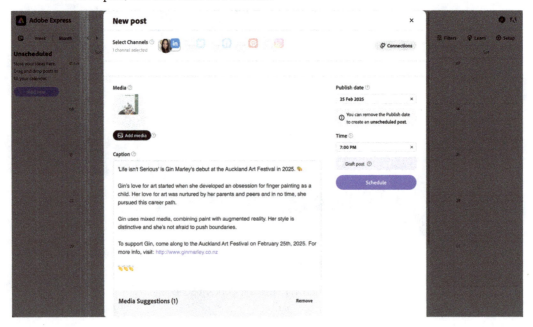

Figure 13.17 – Scheduling your post

16. To view your scheduled post on the Express Content Scheduler, simply navigate to the scheduled publish date on the content calendar.

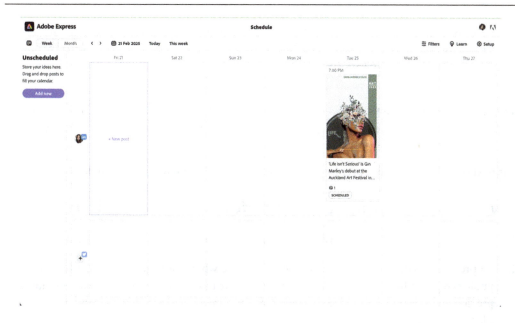

Figure 13.18 – View your scheduled content

17. If you are uncertain about the specific scheduling of the post, simply click and drag the content to the **Unscheduled** tab located on the left side of the page. By doing so, Express enables you to store content here and select a publish date at a later time.

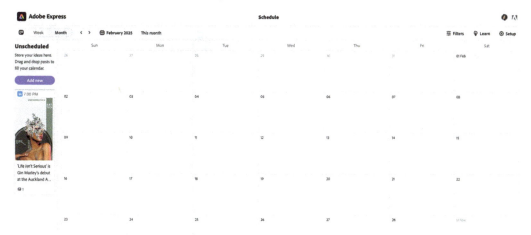

Figure 13.19 – Saving unscheduled posts

18. In the top-right corner, click on **Filters**. This option allows you to filter the post state based on categories such as **Draft, Scheduled, Published, and Failed**. Additionally, you can also filter the channels by selecting specific platforms such as LinkedIn, Twitter, Facebook, Pinterest, and Instagram.

In this section, you acquired the knowledge to schedule a post effectively. You learned how to select content from your Express projects, add captions, connect to your social media accounts, and determine the date and time for publishing.

In the next section, you will discover how to preview your content before a post goes live.

How to preview your content before a post goes live

In this section, you will learn the process of previewing your content before a post is scheduled to go live:

1. To proceed, access the calendar and locate the date when your scheduled post is set to be published. From there, you will have a comprehensive bird's eye view that includes thumbnail previews of the content scheduled for release. You can choose to view the calendar in either **Month** or **Week** format.

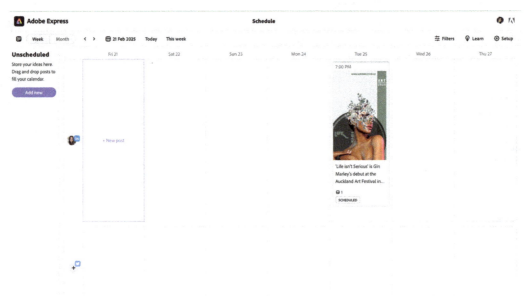

Figure 13.20 – A bird's eye view of the calendar

2. Click on the post you want to preview.

3. Express will open the post, giving you the flexibility to edit details if necessary. To preview the post, simply click on the **Preview** button in the bottom-left corner.

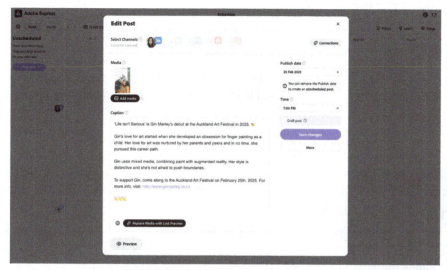

Figure 13.21 – Editing and viewing the scheduled post

4. When you click on the **Preview** button, Express will present you with a preview of your scheduled post.

Figure 13.22 – Previewing your scheduled post

In this section, you acquired knowledge on how to access the live preview of your scheduled post. Express also gives you the flexibility to edit the post. The live preview enables you to visualize how your content will be displayed on the chosen social media platform, providing you with peace of mind and confidence in your publishing decisions.

Summary

In this chapter, you discovered the simplicity of scheduling your social media posts. You can effortlessly upload your content or select a project created in Express. By adding a caption, choosing your social media channel, and specifying the desired publish date and time, you can effectively plan your content in advance. This strategic approach allows better content creation, as you now have a well-defined plan.

With this seamless automation, you no longer have to manually upload content to every single social media platform. Instead, you can efficiently batch your content in Express and let the Content Scheduler handle the heavy lifting for you. You can sit back, relax, and watch as your scheduled posts are seamlessly published.

Congratulations on completing this learning journey with Adobe Express! You have successfully acquired new skills and should be confident with creating visually impressive social media graphics, attention-grabbing animations, beautiful web pages, and captivating videos. Adobe Express provides an intuitive platform that makes learning and implementing these skills a breeze. Take pride in what you have accomplished throughout each chapter of this book. With the newfound knowledge you have gained from this book, you are now fully equipped to produce engaging content that stands out on social media and beyond.

Index

W

Packtpub.com

Subscribe to our online digital library for full access to over 7,000 books and videos, as well as industry leading tools to help you plan your personal development and advance your career. For more information, please visit our website.

Why subscribe?

- Spend less time learning and more time coding with practical eBooks and Videos from over 4,000 industry professionals

- Improve your learning with Skill Plans built especially for you

- Get a free eBook or video every month

- Fully searchable for easy access to vital information

- Copy and paste, print, and bookmark content

Did you know that Packt offers eBook versions of every book published, with PDF and ePub files available? You can upgrade to the eBook version at packtpub.com and as a print book customer, you are entitled to a discount on the eBook copy. Get in touch with us at customercare@packtpub.com for more details.

At www.packtpub.com, you can also read a collection of free technical articles, sign up for a range of free newsletters, and receive exclusive discounts and offers on Packt books and eBooks.

Other Books You May Enjoy

If you enjoyed this book, you may be interested in these other books by Packt:

Edit Like a Pro with iMovie

Regit

ISBN: 9781803238906

- Soak up the principles of editing—coherence, conciseness, and adding meaning
- Use iMovie's Magic Movie and Storyboard tools to create simple, themed videos
- Navigate movie mode for iOS, iPadOS, and macOS and create videos without a template.
- Improve the viewing experience with overlays and use keyframes for smooth animations.

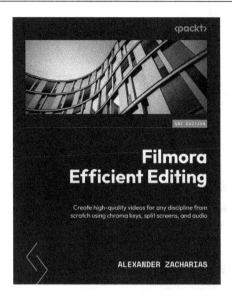

Filmora Efficient Editing

Alexander Zacharias

ISBN: 9781801814201

- Navigate Filmora's interface with ease
- Add and manipulate audio using audio tracks
- Create high-quality professional videos with advanced features in Filmora
- Use split screens and Chroma keys to create movie magic
- Create a gaming video and add humor to it

Packt is searching for authors like you

If you're interested in becoming an author for Packt, please visit `authors.packtpub.com` and apply today. We have worked with thousands of developers and tech professionals, just like you, to help them share their insight with the global tech community. You can make a general application, apply for a specific hot topic that we are recruiting an author for, or submit your own idea.

Hi!

Rosie here, the author of *Express Your Creativity with Adobe Express*. I hope you've been having a fantastic time delving into the world of creativity with this book. My ultimate goal was to make it an enjoyable and valuable resource for enhancing your creative skills, leaving you feeling empowered and inspired to effortlessly create visually stunning content.

It would really help us (and other potential readers!) if you could leave a review on Amazon sharing your thoughts on *Express Your Creativity with Adobe Express*.

Go to the link below or scan the QR code to leave your review:

`https://packt.link/r/1803237740`

Your review will help us to understand what's worked well in this book, and what could be improved upon for future editions, so it really is appreciated.

Best wishes,

Rosie Sue

Download a free PDF copy of this book

Thanks for purchasing this book!

Do you like to read on the go but are unable to carry your print books everywhere?

Is your eBook purchase not compatible with the device of your choice?

Don't worry, now with every Packt book you get a DRM-free PDF version of that book at no cost.

Read anywhere, any place, on any device. Search, copy, and paste code from your favorite technical books directly into your application.

The perks don't stop there, you can get exclusive access to discounts, newsletters, and great free content in your inbox daily

Follow these simple steps to get the benefits:

1. Scan the QR code or visit the link below

https://packt.link/free-ebook/9781803237749

2. Submit your proof of purchase

3. That's it! We'll send your free PDF and other benefits to your email directly

www.ingramcontent.com/pod-product-compliance
Lightning Source LLC
Chambersburg PA
CBHW062058050326
40690CB00016B/3142